UFO HOW-TO "THE BASICS"

SECOND EDITION

Want to talk to the author? Please address all questions and concerns to:

AUTHOR@UFOHOWTO.COM

Or visit us at

WWW.UFOHOWTO.COM

The Basics -
The UFO How-To Sampler

ISBN: 978-1-312-21901-4

TABLE OF CONTENTS

WARNING:

EXTREME DANGER!

THE PATENTS CONTAINED HEREIN USE:

1) HIGH VOLTAGES
2) POWERFUL ELECTROMAGNETIC FIELDS
3) HIGH BURN TEMPERATURES
4) POTENTIAL RADIOACTIVITY

DO NOT ATTEMPT TO BUILD THESE DEVICES WITHOUT MAXIMUM SAFETY PRECAUTIONS!

INTRODUCTION

LET ME START BY SAYING, I AM NOT A DEBUNKER.

THE QUESTION PRESENTED HERE IS: DOES MANKIND POSSESS UFO TECHNOLOGY?

THE ANSWER IS A RESOUNDING **YES**!

I HAVE BEEN ASKED, "IF YOU WERE GOING TO BUILD A UFO-STYLE CRAFT, WHERE WOULD YOU START?"

MY ANSWER LAY IN THESE FOLLOWING PAGES.

THIS BOOK WAS WRITTEN IN RESPONSE TO THE NUMBER OF REQUESTS TO SIMPLIFY THE UFO PHENOMENON. WE WILL BREAK DOWN THE SCIENCE CONTAINED IN THE 3000 PAGES OF THE UFO HOW-TO SERIES WITH SINGLE EXAMPLES AND WITH SIMPLIFIED EXPLANATIONS.

A MAGNETOHYDRODYNAMIC CRAFT PATENT FILED IN 1964, SHOWN HERE AT PAGE 3, WAS FOR A HEAVIER-THAN-AIR FLYING CRAFT SURROUNDED BY A PLASMA.

PLASMA, THE FOURTH STATE OF MATTER, CAN BE CREATED BY HIGH VOLTAGE INTERACTIONS WITH GASES, AND CONTAINED BY POWERFUL MAGNETIC AND/OR ELECTROSTATIC FIELDS.

A PLASMA SPHERE AROUND AN MHD CRAFT WILL MAKE THAT CRAFT APPEAR AS A GLOWING BALL OF LIGHT. THAT SAME PLASMA ABSORBS RADIO WAVES AND MAKES THE CRAFT INVISIBLE TO RADAR.

DOES THIS SOUND LIKE ANY UFO SIGHTING YOU'VE EVER HEARD OF BEFORE?

YOU CAN LEARN MORE ABOUT THIS TOPIC IN VOLUME IV OF THE UFO HOW-TO SERIES: "MAGNETOHYDRODYNAMICS."

May 30, 1967 J. F. KING, JR **3,322,374**

MAGNETOHYDRODYNAMIC PROPULSION APPARATUS

Filed Sept. 30, 1964 2 Sheets-Sheet 1

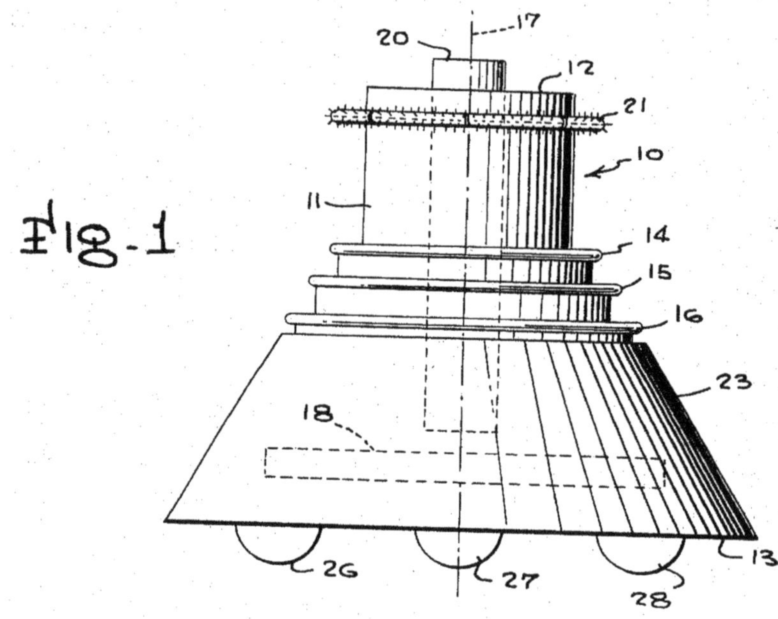

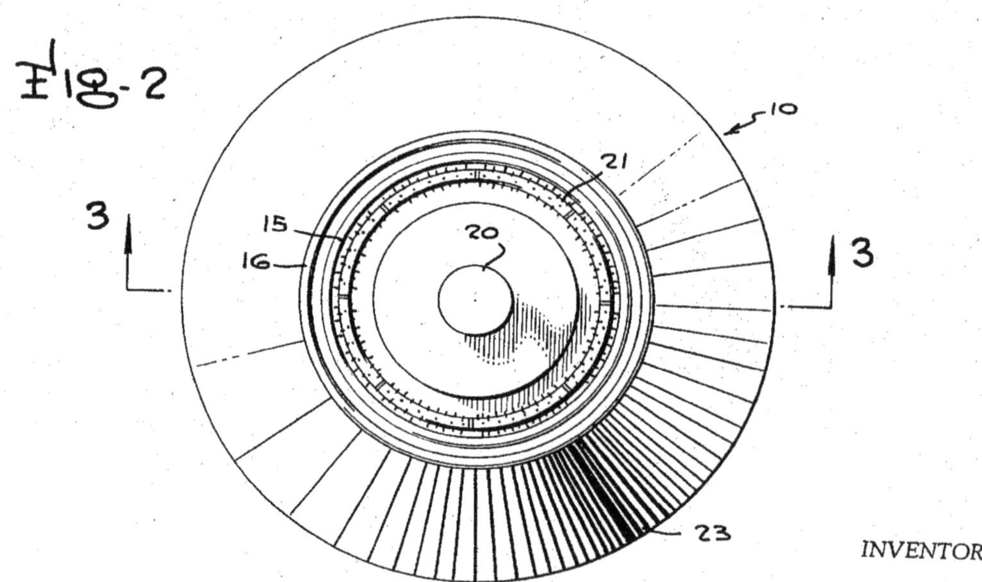

INVENTOR

JAMES F. KING, JR.

BY

Mason, Fenwick & Lawrence
ATTORNEYS

May 30, 1967 J. F. KING, JR 3,322,374

MAGNETOHYDRODYNAMIC PROPULSION APPARATUS

Filed Sept. 30, 1964 2 Sheets-Sheet 2

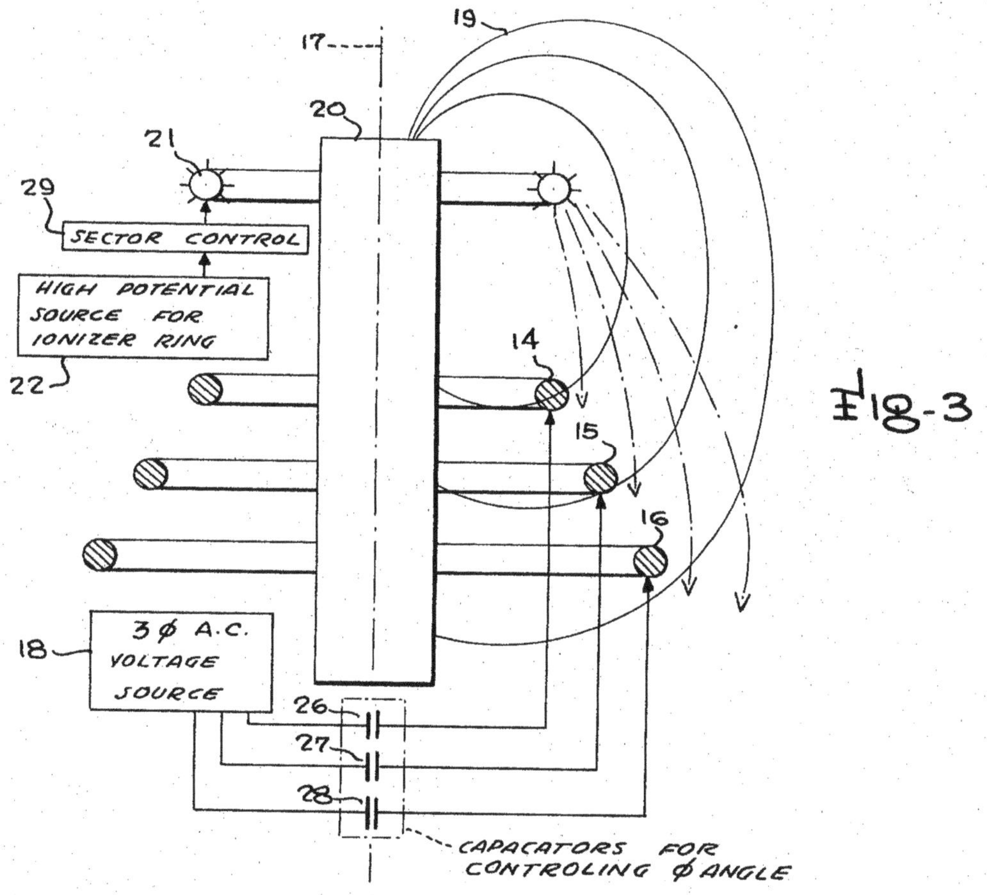

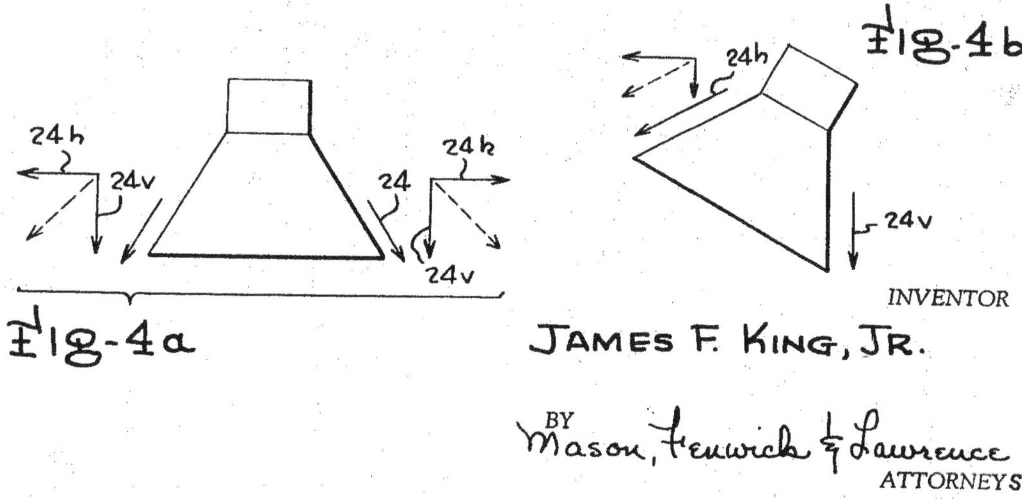

INVENTOR
JAMES F. KING, JR.

BY
Mason, Fenwick & Lawrence
ATTORNEYS

United States Patent Office 3,322,374
Patented May 30, 1967

1

3,322,374
MAGNETOHYDRODYNAMIC PROPULSION APPARATUS

James F. King, Jr., 925 Goodwood Road, Winston-Salem, N.C. 27106
Filed Sept. 30, 1964, Ser. No. 400,456
9 Claims. (Cl. 244—62)

The present invention relates in general to craft propelled by magnetohydrodynamic effects and methods of propulsion and control thereof, and more particularly to heavier-than-air craft which are propelled by interaction of magnetic fields upon electrically conductive fluids such as plasma, surrounding the craft.

The technological field of magnetohydrodynamics, frequently referred to as MHD, is concerned with the study of dynamic effects of magnetic fields upon electrically conducting fluids, a prime example of which is plasma. The term "plasma," has been variously defined as a space charge neutralized ion cloud containing substantially equal numbers of positive ions and negative electrons, or any mixture of particles, some of which are charged, whose spatial dimension exceeds the Debye length and where the percentage of the mixture that is ionized contains an approximately equal number of positive and negative particles so that the overall aggregate is electrically neutral. As used in the present discussion, the term "plasma" is intended to described a gas or electrolyte which in addition to meeting the criteria just given is in such a state of ionization that it becomes conductive enough to be affected by magnetic fields. That is to say, such an electrically conductive fluid medium containing charged particles is sufficiently conductive so that electric currents in the nature of eddy currents may be induced in the fluid medium by magnetic fields by the phenomena known as "mutual induction."

An object of the present invention is the provision of a novel method and apparatus for propulsion of craft which relies upon interaction of magnetic fields produced by electrical currents in conductors on the craft with a surrounding electrically conductive environment or medium to produce reaction thrust.

Another object of the present invention is the provision of a novel propulsion method and apparatus for heavier-than-air craft surrounded by a plasma or ionized field produced by the craft.

Another object of the present invention is the provision of a heavier-than-air craft having self-contained means for generating an ionized or plasma field in air surrounding the craft and means for generating a polyphase excited moving magnetic field of such character that currents are induced in the surrounding ionized or plasma field which constitutes a mobile fluid conductor and the conductor medium is propelled by the moving magnetic fields to produce reactive thrust for propelling the craft.

Yet another object of the present invention is the provision of propulsion apparatus for a craft of the character described in the preceding paragraph arranged in such a way as to permit direction control and impart inherent stability to the craft similar to that attained with dihedral wing arrangements.

Other objects, advantages and capabilities of the invention will become apparent from the ensuing detailed description and accompanying drawings.

Heretofore, arrangements have been disclosed for propulsion of craft by establishing high electrical D.C. potentials between spaced conductors or electrodes and thereby generating ions or charged particles which are electrically attracted in a selected direction and through collision with air molecules create a propulsive force.

2

Examples of such arrangements may be found in U.S. Patent No. 3,120,363 to Hagan and U.S. Patents No. 2,949,550, No. 3,018,394 and No. 3,022,430 to T. T. Brown. These devices, however, rely on the generation and migration of ions by means of the high D.C. potentials between the spaced electrodes providing an action force at least predominantly in a selected direction and a consequent reaction force on the device in the opposite direction pursuant to Newton's third law of motion.

Another arrangement which was proposed in United Kingdom Patent No. 830,816, published Mar. 23, 1960, involved the provision of a "solenoid" of one or more superconductor rings in a space craft through which a powerful current is caused to flow to induce a magnetic field, together with means to ionize the region surrounding the craft and laterally spaced electrodes between which D.C. potential is maintained to pass current through the ionized region along current paths at right angles to the desired direction of travel. The magnetic field in this instance is indicated to be stationary with respect to the field-producing conductor and thus produced by a D.C. voltage, and thrust is said to be produced by the action of the magnetic field to accelerate the ions in the current paths between the electrodes in a direction at right angles to the current paths and opposite to the desired direction of travel. However, nothing contained in this patent suggests that the "solenoid" or magnetic field generating conductive loop of the patented vehicle creates anything other than a stationary field similar to that of a conventional D.C. solenoid winding.

The present invention, instead of relying on such a system of thrust generation, involves the production of a plasma or electrically conductive field surrounding the craft, if the craft is not emersed in such an electrically conductive surrounding medium, and generation of a high intensity moving polyphase excited magnetic field about the craft produced by alternating current voltages in such a manner as to induce eddy currents in the plasma and causing the field to travel between two spaced points in a direction opposite to the desired direction of travel. Since under the Lenz's law, the current set up by an induced voltage always opposes the motion or charge in current which produced it, it follows that the currents (and thus eddy currents) set up in a conductor such as in the ionized or plasma field, located in the moving magnetic field opposes the motion of the field. The consequent motion of the conductor medium produces a reaction force to propel the craft in the desired direction. This is achieved by providing a series of electrically conductive inductive driving rings or coils spaced along a reference axis along which the craft is to be propelled. The driving rings are excited by polyphase high intensity A.C. currents generated by a suitable source within the craft and applied to the series of rings in a manner known in the art, for example in linear motor catapults, electric guns, electrohydraulic pumps for conducting liquids and the like, to produce a moving magnetic field of varying intensity wherein the point of maximum flux density travels progressively from the leading end or top of the craft to the trailing end or bottom of the craft relative to the reference axis, then switches back to the leading end and repeats the travel. If the craft is not designed solely for travel in a natural environment of highly electrically conductive medium, an air ionizer is provided at the leading end or top of the craft to highly ionize the air surrounding the craft and provide a surrounding field or cloud in which eddy currents may be induced by the magnetic field generated by the series of driving rings. Such a traveling field, when acting upon the surrounding conductive medium serving as a conductor, generates or induces eddy current therein and this induced current flowing between finite particles of atmosphere making up the conductor coacts with the flux of the field to set up a force on the conductor tending to cause the conductor to move with the traveling field. A powerful thrust is exerted on the craft in the desired direction of travel by the oppositely directed travel of the conductor (i.e. the ionized or plasma medium). The ionizer in effect produces a hollow electrically conductive tube or annulus of highly ionized air or plasma surrounding the craft through which the craft is to travel and the rearwardly traveling primary magnetic field generated by the driving rings induces the eddy currents in this conductive tube causing the "tube" of conductive medium to be thrust rearwardly and thereby propel the craft along the tube. Stability is achieved by forming the driving rings of progressively increasing diameter progressing downwardly or toward the trailing end or bottom of the craft, and shaping the craft to direct the surrounding air and thus the ionized field or zone generally outwardly and downwardly to produce effects similar to those of dihedral in winged aircraft.

An illustrative embodiment of a craft incorporating the principles of the present invention is shown in the accompanying drawings in somewhat diagrammatic fashion, in which:

FIGURE 1 is a side elevation of a craft embodying the present invention;

FIGURE 2 is a top view thereof;

FIGURE 3 is a diagrammatic skeleton view of the primary magnetic field generating and ionizer components of the craft arranged in their relative positions to each other to illustrate the principles of operation thereof, taken along the plane 3—3 of FIGURE 2; and

FIGURES 4*a* and 4*b* are diagrammatic illustrations of the craft at vertical and tilt attitudes, respectively, with accompanying illustrations of vector forces to show the stabilizing effects of the configuration.

Referring to the drawings, the heavier-than-air craft, indicated generally by the reference character **10**, comprises an air frame or fuselage **11** providing support for the various compartments of the craft and the load to be transported, having a leading end or top **12** and a trailing end or bottom **13**. At an intermediate location between the top **12** and bottom **13** are a series of polyphase driving windings or coils **14, 15** and **16** spaced axially along the craft concentric with the center or reference axis **17** thereof, with the relatively trailing rings **15** and **16** located toward the trailing end **13** being of progressively greater diameter to define a generally conical arrangement diverging toward the trailing end **13**. To generate a magnetic field of sufficient strength, a very heavy current must flow through a single conductor or a more moderate current through many turns. Since a single turn is more efficient, I prefer to provide a conductor in the form of a single turn closed driving ring for each of the coils **14, 15** and **16**, and apply three phase A.C. voltage to the driving rings **14, 15** and **16** from a suitable source **18** with proper connections to produce a traveling magnetic field, indicated by the lines **19** in FIGURE 3, wherein the point of maximum flux density progresses from the top to the bottom of the stack of driving rings, and then switches back to the top and progresses again to the bottom. Preferably a permeable pole piece **20** extends along the center axis **17** of the craft to convert the driving rings **14, 15, 16** from an air core to a ferrite core coil system to increase the flux density. The art of connecting the polyphase supply voltage to such rings or coils is well understood as exemplified by numerous prior patents directed to linear-motor catapults, electrohydraulic pumps for electrically conductive fluids, electric projectile propelling apparatus, and the like, typical of which are U.S. patents Nos. 2,112,264 to Bowles, 2,428,570 to Jones, 2,920,571 to Fenemore et al., 3,008,418 to Blake, 3,005,-116 to Reece, 3,135,879 to Baumann and 2,214,297 to Ferry.

To provide maximum magnetic field strength with minimum coil weight, the driving rings **14, 15, 16** are preferably constructed of conductive metal material maintained in a superconductive state, for example by refrigeration or cooling equipment carried by the craft.

To provide a surrounding medium in the form of a flowable mass of electrically conductive properties to serve as the "conductor" in which the traveling primary field produced by the driving rings **14, 15, 16** can induce eddy currents by mutual induction, an ionizer coil or ring **21** is supported at the top **12** of the craft supplied by a suitable high voltage source **22**, to cause the surrounding air to be ionized as fast as it enters the magnetic field of the driving rings **14, 15, 16**. It is recognized that as air becomes more and more highly ionized, it becomes a progressively better conductor until at a condition which supports disruptive breakdown, it is an almost perfect conductor. The ionizer **21** at the top or leading end of the craft therefore continuously ionizes the air at the leading end of the craft, providing a surrounding annulus of highly conductive medium or plasma in which the eddy currents can be induced by the driving rings **14, 15, 16**. Particle repulsion as well as the repulsion of the self-induced secondary magnetic fields established by the eddy currents in the surrounding ion cloud or plasma will spread the ionized annulus to the extent of the magnetic lines of force away from the craft before they reach the level of maximum flux density as they relatively progress toward the trailing end of the craft. The movement of the driving magnetic field downwardly, as previously mentioned, set up forces on the conductor medium causing the ionized or plasma medium to travel with the driving magnetic field and thus downwardly about the craft.

Preferably, means are also provided such as the frusto-conical air-directing skirt **23** disposed just below the set of driving rings **14, 15, 16** to direct the ionized air, which moves relative to the craft toward the trailing end **13**, in outwardly inclined paths at an angle relative to the reference axis **17** to more nearly shape the field of ions to fit the magnetic field produced by the similarly outwardly and downwardly inclined paths of the driving rings **14, 15, 16**. This inclining of the ionized air flow paths outwardly and downwardly achieves improved stability for the craft, as will be apparent from an inspection of FIGURES 4*a* and 4*b*. With the craft **10** in the vertical attitude illustrated in FIGURE 4*a*, the ionized air flow as designated by the arrows **24** lies in a downwardly divergent conical path due to the effects of the air directing skirt **23**. These forces or vectors all resolve into a downwardly directed vertical component $\mathbf{24}_v$ and a radially outwardly directed horizontal component **24**, the latter occurring in a circle so that they cancel out each other and only the vertical component remains. Should the craft **10** tilt as illustrated in FIGURE 4*b*, the ionized air force vector $\mathbf{25}_L$ for the lower side of the craft is all down while the vector $\mathbf{25}_H$ for the higher opposite side is near horizontal, producing a couple which tends to return the craft to its proper vertical attitude. It will be appreciated that the force vectors shown represent the forces exerted on the eddy-current-conducting ionized air cloud (i.e. the conductor) by the downwardly traveling primary magnetic field produced by the driving rings **14, 15** and **16**.

While the phase angle between current and voltage in electric motors is not great due to partial phase shift from inductive reactance by circuit resistance, and can tolerate the consequent efficiency loss, the use of superconductor driving rings or coils **14, 15, 16** in the craft of the present invention would substantially eliminate resistance and provide substantially pure inductance, whereby the voltage would lag the current by approxiamtely 90° and no magnetism would result from the current in the coils. To shift the voltage back in phase with the current, capacitance is added to the circuit, preferably by capacitors sufficient to provide capacitive reactance equal to the inductive reactance and thereby place the circuit in resonance with a power factor of unity. This may be accomplished by providing large condensers **26, 27, 28** in series resonant circuits with the respective driving rings **14, 15, 16.**

Directional and attitude control of the craft may be effected either by control of the ionization in various radial directions or quadrants, or by distorting the shape of the magnetic field produced by the driving rings **14, 15, 16** or by other control expedients. For example, the ionizer ring **21** may, in a simple form, be a ring of spikes with alternate spikes at ground potential and a very substantial potential difference existing between the spikes so as to effect a high degree of ionization between rows of alternate spikes. The spiked ring **21** may be broken up along its circumference into a plurality of independently controllable arcuate segments, and the potential applied to the arcuate segments through a sector control **29** which moderates the applied potential to set the degree of ionization in each quadrant. By this means, the ionization can be altered to maintain constant vertical force components or lift or vary them to effect controlled tilt, while the normally zeroed out horizontal force components can be upset so that the craft has a powerful horizontal force component in the desired direction. This may be coordinated with control of total power input to the driving rings **14, 15, 16** to maintain the correct proportions of lift, tilt and forward speed for controlled flight providing control functions similar to collective pitch control in rotary wing aircraft.

Lateral directional control may also be achieved by installing a plurality of closed loop coils in the skirt of the craft below the primary coils. These directional control coils may lie in the plane of the skirt and take the form of a series of circles or regular loops which are tangent to one another around the perimeter of the craft in the skirt region. These closed loop coils would be controllable as to whether they are closed or open, for example, by remotely controlled shorting strips or conductive bridges, so that if they are closed, they form a short circuited secondary turn which, by resistance, would absorb power from the main driving coils and tend to shade the developed magnetic field at that point. If the closed circuit of this coil is broken, no current would be able to flow and no power would be absorbed. By thus decreasing the field strength in any quadrant, one would be able to tilt the craft and achieve lateral displacement.

To effect control over the amount of thrust generated, one could increase power to either the driving coils to the ionizer ring, or if desired, one may change the frequency of the phased alternating current voltages supplied to the driving coils to increase or decreases the speed of travel of the travelling magnetic field along the driving rings.

In the event the craft is to have the capability of travel in a medium or region having insufficient air or other ionizable matter, an ion source or the like would also have to be carried by the craft to provide its own source of ions or electrically conductive matter for seeding the surrounding region and producing the necessary plasma or electrically conductive annulus in which the eddy currents may be induced to provide propulsive thrust. If the craft is to travel in a fluid medium which is naturally electrically conductive, such as sea water or some other electrolyte, the medium already has the capability of conducting eddy currents and the ionizer may be dispensed with or may be used merely to improve conductivity of the naturally conductive medium.

While but one preferred example of the present invention has been particularly shown and described, it is apparent that various modifications may be made therein within the spirit and scope of the invention, and it is desired, therefore, that only such limitations be placed on

the invention as are imposed by the prior art and set forth in the appended claims.

What is claimed is:

1. A magnetohydrodynamic effects craft adapted to travel through a surrounding fluid medium wherein the zone of the medium adjacent the craft is responsive to magnetic fields to conduct mutually induced eddy currents, comprising a plurality of electrically conductive coil means capable of producing magnetic fields surrounding the craft upon conduction of electric current therethrough arranged serially along an axis of the craft paralleling a desired direction of travel, means for applying polyphase alternating current electric voltages to said coil means to produce a plurality of magnetic fields fluctuating responsive to alternating currents in said coil means which are phased and magnetically superimposed to produce a collective driving magnetic field surrounding said craft whose point of maximum flux density repetitively progresses from a leading end position relative to said direction of travel to a trailing end position and switches back to said leading end position, said driving magnetic field being of sufficient strength to induce eddy currents in the adjacent zone of said fluid medium and thereby constitute the fluid medium an equivalent electrical conductor interacting with said driving magnetic field to be driven in the direction of travel of the driving magnetic field and effect propulsion of said craft in the selected direction of travel.

2. A heavier-than-air craft adapted to be propelled through a fluid medium by magnetohydrodynamic effects comprising a fuselage having a top end and a bottom end spaced along a central reference axis, means adjacent the top end of said craft for ionizing the fluid medium surrounding the craft, a plurality of inductive driving coils of electrical conductors arranged concentrically of said reference axis and located at a plurality of selected spaced positions along said axis intermediate said top and bottom ends, and means for applying polyphase alternating current voltages to said coils to produce electric currents therein having selected phase relations to each other establishing a magnetic field about each coil fluctuating about its respective coil in accordance with the alternating electric currents flowing therein, which fields are phased and magnetically superimposed to produce a collective driving magnetic field whose point of maximum flux density repetitively progresses from the topmost coil to the bottommost coil and switches back to the top-most coil, whereby said driving magnetic field induces eddy currents in the ionized surrounding medium reacting with said driving magnetic field to drive the ionized surrounding medium in the direction of travel of the driving magnetic field and propel the craft upwardly in the direction of said reference axis.

3. A heavier-than-air craft adapted to be propelled through a fluid medium capable of being ionized to a condition of high electrical conductivity by magnetohydrodynamic effects comprising a fuselage having a top end and a bottom end spaced along a central reference axis, means adjacent the top end of said craft for ionizing the fluid medium surrounding the craft to produce an ionized annulus of high electrically conductive fluid medium surrounding the craft and extending axially toward the bottom end thereof, an array of inductive driving rings formed of electrical conductors arranged concentrically of said reference axis and located at selected spaced positions along said axis intermediate said top and bottom ends, and means for applying polyphase alternating current voltages to said rings to produce electric currents therein having selected phase relations to each other establishing a magnetic field about each ring fluctuating about its respective ring in accordance with the alternating electric currents flowing therein, which fields are phased and magnetically superimposed to produce a collective driving magnetic field whose point of maximum flux density repetitively progresses from the

3,322,374

7

top to the bottom of said array and switches back to the top of the array, whereby said driving magnetic field induces eddy currents in the ionized surrounding medium reacting with said driving magnetic field to drive the ionized surrounding medium in the direction of travel of the driving magnetic field and propel the craft upwardly in the direction of said reference axis.

4. In a heavier-than-air craft as defined in claim 3, each of said driving rings being superconductor rings.

5. A heavier-than-air craft adapted to be propelled through a fluid medium by magnetohydrodynamic effects comprising a fuselage having a top end and a bottom end spaced along a central reference axis, means adjacent the top end of said craft for ionizing the fluid medium surrounding the craft, a plurality of inductive driving coils of electrical conductors arranged concentrically of said reference axis and located at a plurality of selected spaced positions along said axis intermediate said top and bottom ends, and means for applying polyphase alternating current voltages to said coils to produce electric currents therein establishing a moving driving magnetic field which repetitively progresses from the top-most coil to the bottom-most coil and switches back to the top-most coil, whereby said moving magnetic field induces eddy currents in the ionized surrounding medium reacting with said driving magnetic field to drive the ionized surrounding medium in the direction of travel of the magnetic field and propel the craft upwardly in the direction of said reference axis, said fuselage having fluid medium deflecting skirt means of generally truncated conical configuration diverging toward said bottom end between said coils and said bottom end to deflect the ionized medium surrounding the craft in outwardly and downwardly inclined paths relative to said reference axis during relative movement of the craft along said axis for developing forces upon interaction of said driving magnetic field with the ionized fluid medium tending to stabilize the attitude of the craft relative to said axis.

6. A heavier-than-air craft adapted to be propelled through a fluid medium capable of being ionized to a condition of high electrical conductivity by magnetohydrodynamic effects comprising a fuselage having a top end and a bottom end spaced along a central axis, an ionizer ring adjacent the top end of said craft, means for applying a high electrical potential to said ionizer ring for ionizing the fluid medium surrounding the craft to produce an ionized annulus of high electrically conductive fluid medium surrounding the craft and extending axially toward the bottom end thereof, an array of inductive driving rings formed of electrical conductors arranged concentrically of said reference axis and located at selected spaced positions along said axis intermediate said top and bottom ends, and means for applying polyphase alternating current voltages to said rings to produce electric currents therein having selected phase relations to each other establishing a magnetic field about each ring fluctuating about its respective ring in accordance with the alternating electric currents flowing therein, which fields are phased and magnetically superimposed to produce a collective driving magnet field whose point of maximum flux density repetitively progresses from the top to the bottom of said array and switches back to the top of the array, whereby said driving magnetic field induces eddy currents in the ionized surrounding medium reacting with said driving magnetic field to drive the ionized surrounding medium in the direction of travel of the driving magnetic field and propel the craft upwardly in the direction of said reference axis.

7. A heavier-than-air craft adapted to be propelled through a fluid medium capable of being ionized to a condition of high electrical conductivity by magnetohydrodynamic effects comprising a fuselage having a top end and a bottom end spaced along a central axis, an ionizer ring adjacent the top end of said craft, means for applying a high electrical potential to said ionizer ring for ionizing

8

the fluid medium surrounding the craft to produce an ionized annulus of high electrically conductive fluid medium surrounding the craft and extending axially toward the bottom end thereof, an array of inductive driving rings formed of electrical conductors arranged concentrically of said reference axis and located at selected spaced positions along said axis intermediate said top and bottom ends, and means for applying polyphase alternating current voltages to said rings to produce electric currents therein, establishing a moving driving magnetic field which repetitively progresses from the top to the bottom of said array and switches back to the top of the array, whereby said moving magnetic field induces eddy currents in the ionized surrounding medium reacting with said driving magnetic field to drive the ionized surrounding medium in the direction of travel of the magnetic field and propel the craft upwardly in the direction of said reference axis, said means for applying electrical potential to said ionizer ring including means for controlling the distribution of potential about the circumference of said ionizer ring to vary the potential at selected arcuate portions of the ionizer ring and thereby alter the ionization in various azimuthal zones surrounding the craft to guide the craft.

8. A heavier-than-air craft adapted to be propelled through a fluid medium capable of being ionized to a condition of high electrical conductivity by magnetohydrodynamic effects comprising a fuselage having a top end and a bottom end spaced along a central axis, an ionizer ring adjacent the top end of said craft, means for applying a high electrical potential to said ionizer ring for ionizing the fluid medium surrounding the craft to produce an ionized annulus of high electrically conductive fluid medium surrounding the craft and extending axially toward the bottom end thereof, an array of inductive driving rings formed of electrical conductors arranged concentrically of said reference axis and located at selected spaced positions along said axis intermediate said top and bottom ends, and means for applying polyphase alternating current voltages to said rings to produce electric currents therein, establishing a moving driving magnetic field which repetitively progresses from the top to the bottom of said array and switches back to the top of the array, whereby said moving magnetic field induces eddy currents in the ionized surrounding medium reacting with said driving magnetic field to drive the ionized surrounding medium in the direction of travel of the magnetic field and propel the craft upwardly in the direction of said reference axis, said ionizer ring comprising a plurality of electrically independent, circumferentially spaced arcuate sectors, said means for applying potential to said ionizer ring including sector control means for selectively varying the relative potential applied to said sectors to vary the ionization of the surrounding fluid medium in selected radial directions relative to the reference axis to guide the craft in selected directions.

9. A heavier-than-air craft adapted to be propelled through a fluid medium capable of being ionized to a condition of high electrical conductivity by magnetohydrodynamic effects comprising a fuselage having a top end and a bottom end spaced along a central reference axis, means adjacent the top end of said craft by ionizing the fluid medium surrounding the craft to produce an ionized annulus of high electrically conductive fluid medium surrounding the craft and extending axially toward the bottom end thereof, an array of inductive driving rings formed of electrical conductors arranged concentrically of said reference axis and located at selected spaced positions along said axis intermediate said top and bottom ends, the successive individual driving rings being of progressively increasing diameter progressing from the top end to the bottom end of said craft, and means for applying polyphase alternating current voltages to said rings to produce electric currents therein having selected phase relations to each other establishing a magnetic field about each ring fluctuating about its respective ring in

9

accordance with the alternating electric currents flowing therein, which fields are phased and magnetically superimposed to produce a collective driving magnetic field whose point of maximum flux density repetitively progresses from the top to the bottom of said array and switches back to the top of the array, whereby said driving magnetic field induces eddy currents in the ionized surrounding medium reacting with said driving magnetic field to drive the ionized surrounding medium in the direction of travel of the driving magnetic field and propel the craft upwardly in the direction of said reference axis.

10

References Cited

UNITED STATES PATENTS

3,071,705	1/1963	Coleman et al.	60—35.5 X
3,138,019	6/1964	Fonda-Bonardi	60—35.5 X
3,150,483	9/1964	Mayfield et al.	60—35.5
3,174,278	3/1965	Barger et al.	60—35.5

FOREIGN PATENTS

830,816	3/1960	Great Britain.

FERGUS S. MIDDLETON, *Primary Examiner.*

MORE POWERFUL THAN ROCKETS

THE PLASMA PROPULSION PATENT FILED IN 1959, SHOWN HERE AT PAGE 12 IS CAPABLE OF THRUST 60 TIMES MORE POWERFUL THAN THAT OF CHEMICAL ROCKETS.

CHEMICAL ROCKET SPEEDS:

GROUND TO LOW EARTH ORBIT	17,000 MPH
GROUND TO EARTH ESCAPE	24,200 MPH
GROUND TO LUNAR ORBIT	25,700 MPH

THE FOLLOWING PATENT, "PLASMA PROPULSION DEVICE" COULD EASILY BE INCORPORATED INTO THE CRAFT DETAILED AT PAGE 4, OR A CRAFT MADE UP OF COMBINING THE PAGE 4 AND PAGE 12 TECHNOLOGIES. THE PLASMA THRUSTERS[1] COULD EASILY ADD ACCELERATION ABILITY THAT COULD PROPEL THE CRAFT ACROSS THE GLOBE IN MERE SECONDS.

YOU CAN LEARN MORE ABOUT PLASMA THRUSTERS AND THE RELATED SCIENCE IN VOLUME III OF THE UFO HOW-TO SERIES: "PLASMA PROPULSION."

[1] INCORPORATING THE ELEMENTS THAT PROPERLY HARNESS THE ELECTROGRAVITIC EFFECT APPEARS TO SHIELD THE HUMAN PILOT AND PASSENGERS FROM THE INCREDIBLE G-FORCES GENERATED.

March 14, 1961 W. L. STARR 2,975,332

PLASMA PROPULSION DEVICE

Filed Dec. 2, 1959

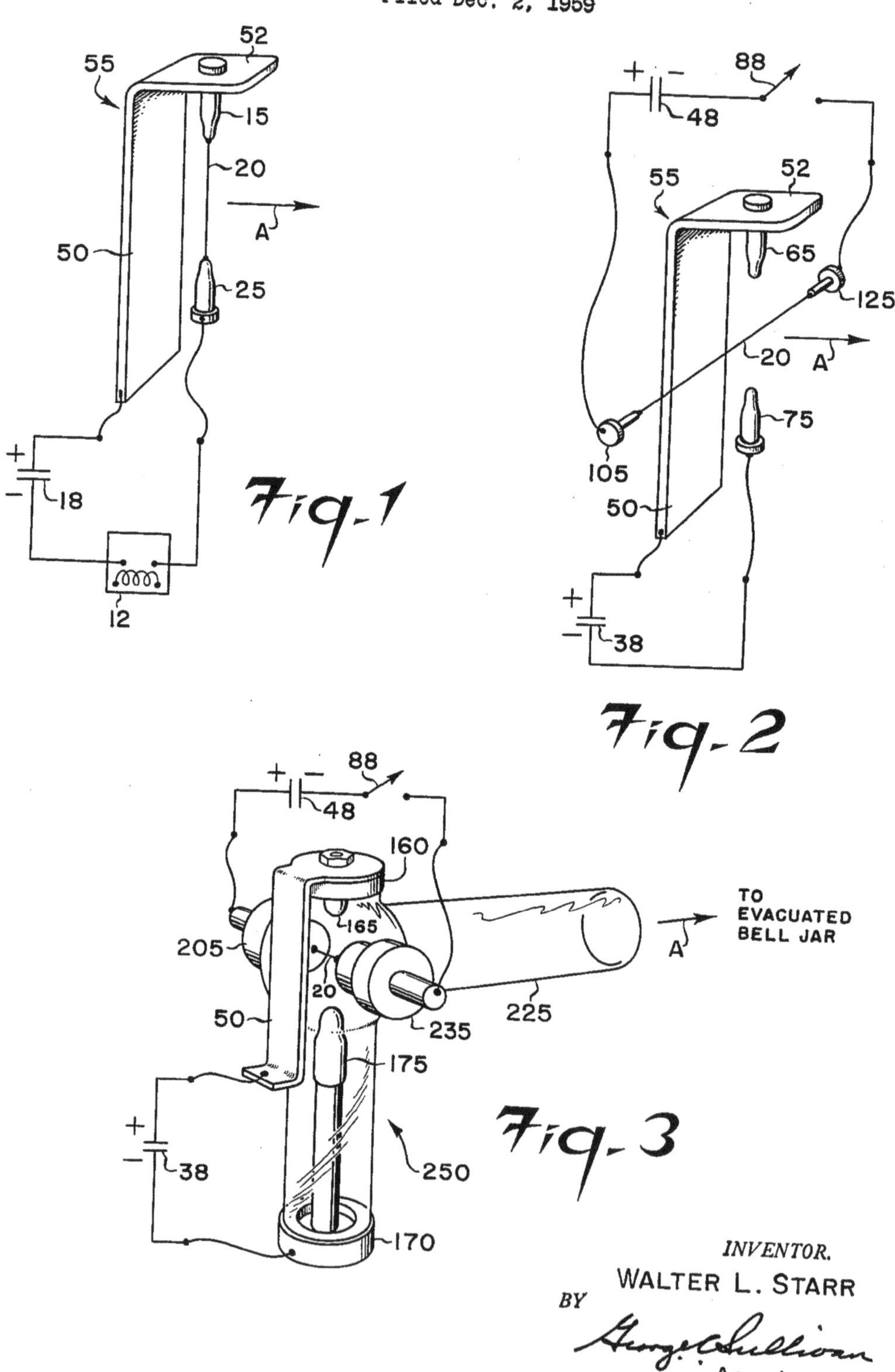

United States Patent Office 2,975,332 Patented Mar. 14, 1961

1

2,975,332

PLASMA PROPULSION DEVICE

Walter L. Starr, Palo Alto, Calif., assignor to Lockheed Aircraft Corporation, Burbank, Calif.

Filed Dec. 2, 1959, Ser. No. 856,821

4 Claims. (Cl. 315—169)

This invention relates generally to propulsion devices, and more particularly to a propulsion device based on plasma acceleration principles and adapted for use in a space environment.

With the advent of space travel, the necessity of providing a propulsion system capable of yielding a very large specific impulse so as to require only a small propellant mass takes on new importance. Accordingly, it is the broad object of this invention to provide a propulsion system adaptable for space propulsion and capable of producing a specific impulse which is very much greater than is now obtainable by current chemical systems.

It is an additional object of this invention to provide improved means and methods for accelerating a plasma to high velocities.

The specific impulse of a propulsion system is defined as the ratio of thrust produced to the rate of expenditure of propellant weight. The propellant velocity is then the ratio of the thrust to the rate of expenditure of propellant mass, thereby being directly proportional to the specific impulse. It can thus be seen that the larger the specific impulse, the less propellant mass that is required, which is a most important consideration for any propulsion system to be used for space travel. The highest specific impulse currently obtainable with operational chemical systems is about 250 seconds. This corresponds to an average propellant velocity of 2.5×10^5 centimeters/second.

It is recognized that a few experiments have been performed in the prior art which are concerned with plasma acceleration techniques. (See, for example, Korneff, Nadig and Bohn, "Plasma Acceleration Experiments," Conference on Extremely High Temperatures, edited by N. Fischer and L. C. Mansur, published by John Wiley and Sons, Inc., New York, 1958.) As far as is known, none of these experiments disclose a plasma acceleration configuration which produces a sufficiently high specific impulse adaptable for space propulsion. Although a suitable plasma acceleration space propulsion system would be highly desirable, progress has been hindered because the interaction between a magnetic field and a plasma is very complex and not fully understood, making it impossible to predict which configuration will produce the highest propellant velocity and thus the highest specific impulse.

As a result of a number of experiments I have conducted, I discovered a novel configuration for a plasma acceleration device which not only provides a high specific impulse, but also, is amazingly simple and well suited for space propulsion. The configuration of the device is such that a combination of forces are caused to act on the plasma (which serves as the propellant) in such a way that it is accelerated to an unexpectedly high velocity.

In a typical embodiment of a propulsion device in accordance with the invention the vapor from an electrically exploded fine wire forms the plasma and is ac-

2

celerated as a result of the interaction between the plasma and the strong forces generated by the novel plasma acceleration configuration. Experimental measurements on such a system indicate that the magnitude of the specific impulse obtained is of the order of 16×10^3 seconds, and the average propellant velocity is on the order of 16×10^6 centimeters/second; this is an increase in the value of the specific impulse and the average propellant velocity of greater than 60 to 1 over that obtainable with current chemical systems. (Comparison figures on prior art plasma acceleration experimental observations are not available).

The specific nature of the invention, as well as other objects, uses, and advantages thereof will clearly appear from the following description and from the accompanying drawing in which:

Figure 1 is a schematic and pictorial diagram of an initial embodiment of a basic plasma propulsion device in accordance with the invention.

Figure 2 is a schematic and pictorial diagram of an improved embodiment of a plasma propulsion system in accordance with the invention.

Figure 3 is a schematic and pictorial diagram of an experimental test model of a propulsion device in accordance with the invention.

Like numerals designate like elements throughout the figures of the drawing.

In Figure 1, a basic propulsion device is shown comprising two electrodes **15** and **25** between which a fine wire **20** is connected. The electrode **15** is electrically and mechanically connected to the bent over portion **52** of a metal right angle member **55** having a vertical portion **50** whose length is preferably at least sufficient to extend the length of the fine wire **20**. This vertical portion **50** will be referred to as the "backstrap" and plays an important part in the propulsion device, as will hereinafter be explained.

A capacitor **18** charged to a high voltage of the order of many thousands of volts is connected in series with the fine wire **20** and the "backstrap" **50** through a conventional type of induction activated air gap switch **12**.

The operation of the Figure 1 propulsion device may now be explained as follows. It must be realized that the operation of the device must take place substantially in a vacuum, as is present in space, and such an environment will be assumed for the Figure 1 system. When the air gap switch **12** is caused to break down, such as by means of an induction coil drive spark, the capacitor **18** discharges its voltage through the fine wire **20** and the "backstrap" **50**. The magnitude of the capacitor **18** and the voltage thereon are chosen so that the discharge of the capacitor **18** vaporizes and heats the fine wire **20** to the plasma state. The large discharge current flowing through the "backstrap" **50** induces a strong magnetic field in the discharge area. The plasma will now be accelerated in the direction of the arrow A shown in Figure 1 as a result of the action of the following three mechanisms: (1) the joule heating of the vapor with the consequent expansion, (2) the repulsive force produced by the magnetic fields resulting from discharge current flow through the two anti-parallel conductors, and (3) the asymmetrical force exerted on a kinked conductor. The kink force in mechanism (3) is a result of a larger magnetic field on the concave side than on the convex side of a kink in a single conductor. The curve produced in the arc discharge by the magnetic field induced by the "backstrap" **50** acts as the kinked conductor.

The action of the three above described mechanisms is believed ideally suited for providing a propulsion system having a high specific impulse, particularly in view of the simplicity of the configuration. However, from a practical viewpoint, it has been found that the particular configuration shown in Figure 1 does not permit the action of these three mechanisms to be used to their fullest extent. Measurements on the propulsion device of Figure 1 indicated that the specific impulse of such a propulsion device is not very much greater than that obtained by current chemical systems. The reason for this appears to be that the plasma, which serves as the propellant, is either in a molten form or a relatively cool vapor when it leaves the electrode area, thereby failing to receive the full effect of the forces produced by the above-described three mechanisms.

Because of the limitations of the Figure 1 device, the configuration shown in Figure 2 was devised. In the Figure 2 device, instead of connecting the fine wire between the electrodes 65 and 75, as was done in the device of Figure 1, the fine wire 20 is disposed at right angles to both the axis of the electrodes 65 and 75 and the desired direction of plasma acceleration indicated at A. This disposition of the fine wire 20 is accomplished by means of the additionally provided electrodes 105 and 125 as shown. The means for vaporizing the wire 20 may thus be provided independently of the discharge circuit.

As shown in Figure 2, vaporization of the wire 20 is obtained by means of a capacitor 48 in series with a switch 88 and the fine wire 20. The size and voltage on the capacitor 48 need only be chosen so that when the switch 88 is closed, current through the wire 20 causes it to become vaporized, forming a plasma which breaks down the electrode gap between the electrodes 65 and 75.

A capacitor 38 in series with the "backstrap" 50 and the electrodes 65 and 75 provides the source of energy for the discharge, as did the capacitor 18 of Figure 1, and is charged to the many thousands of volts required for the discharge. The "backstrap" 50 should preferably have a length at least sufficient to extend the distance between the electrodes 65 and 75. Since the electrodes 65 and 75 are no longer connected together by the wire 20, there is no need for the air gap switch 12 which was used in Figure 1, the free space environment preventing voltage breakdown between the electrodes 65 and 75.

The operation of the three axis configuration of Figure 2 may now be explained as follows. As was the case with the Figure 1 device, operation must take place substantially in a vacuum. When the switch 88 is closed the capacitor 48 discharges through the fine wire 20, causing it to vaporize and form a plasma in the gap between the electrodes 65 and 75. This breaks down the gap and initiates the discharge of the capacitor 38 between the electrodes 65 and 75, causing a large current flow through the "backstrap" 50. As was described in connection with the embodiment of Figure 1, the three mechanisms (1), (2), and (3) act to accelerate the plasma in the direction of the arrow A. Because the wire 20 must be vaporized before the discharge between the electrodes 65 and 75 is initiated, the problem of the plasma not being sufficiently heated, which occurred in the Figure 1 configuration, is not present in the embodiment of Figure 2, thereby permitting full advantage to be taken of the forces produced by these three mechanisms.

The plasma is held in the discharge area by the pinch forces of the discharge until pushed into the direction of the arrow A by both the magnetic forces produced by discharge current flowing through the "backstrap" 50, and kink forces arising from the induced instability of the plasma discharge column. The result is that the plasma is accelerated to an unexpectedly high velocity, producing a specific impulse which is very much larger than is obtainable by current chemical systems.

In Figure 3 an experimental model of a propulsion device in accordance with the three-axis configuration of Figure 2 is illustrated. In order to simulate a space environment, a glass T-shaped electrode tube 250 is employed which is similar to the T-tube used in studies on magnetically driven shocks. The glass tube 250 has a heavy Pyrex wall with an outside diameter of 1-inch and an inside diameter of 11/16-inch. The side arm 225 of the tube 250 is approximately 10 centimeters long. A first electrode 165 depends from a metal member 160 at the top of the tube 250, and a second electrode 175 depends from a metal member 170 at the bottom of the tube 250. The "backstrap" 50 is provided by means of a metal member connected to the top electrode 165 and running adjacent the outside of the tube 250 on the side of the tube 250 farthest from the side arm 225. The "backstrap" extends at least the length of the gap between the electrodes 165 and 175 as shown. Nickel electrodes 205 and 235 to which the fine wire 20 is connected are sealed into opposite ends of the tube 250 so as to dispose the fine wire 20 at right angles to both the axis of the electrodes 165 and 175 and the side arm 225. The capacitor 48 and the switch 88 are connected in series with the fine wire 20 by connections to the nickel electrodes 205 and 235 in any suitable manner. The capacitor 38 is connected to the "backstrap" 50 and to the electrode 175 by means of the extended portion 193 passing through the tube 250. Well known practice is employed to minimize inductive lead effects in order to reduce ringing.

Specific impulse and propellant velocity measurements were taken for the device of Figure 3 in which a 1.1 microfarad capacitor charged to 3,000 volts is used for the capacitor 38 and a 100 microfarad capacitor charged to 3,000 volts is used for the capacitor 48. The fine wire 20 used for these measurements is a 1-millimeter diameter tungsten wire 5 centimeters long. The side arm 221 is connected to a bell jar (not shown), the pressure in the bell jar and the T-tube 250 being kept below 10^{-5} mm. Hg, in order to simulate a space environment. Inside the evacuated bell jar a ballistic pendulum is provided to measure the specific impulse produced by the propulsion test device of Figure 3. The measurements resulted in extrapolated values of specific impulse and propellant velocity of the order of 16×10^3 seconds and 16×10^6 centimeters/second, respectively. In other measurements which were made the fine wire 20 varied in size from 1 to 5 millimeters and a number of different materials were used with comparable results.

In the configuration of Figure 2 and the specific test model thereof illustrated in Figure 3, the fine wire 20 serves as the primary source of propellant. It should be understood, however, that if desired the fine wire 20 could be employed merely for the purpose of initiating the discharge between the main electrodes 65 and 75 in Figure 2, the major part of the propellant being derived from vaporization of material on the electrodes or of other material provided in the vicinity of the discharge area for this purpose. It should also be understood that initiation of the main discharge may be provided in other ways than by using the fine wire 20, as will be evident to those skilled in the art. For example, small button probes are available which when activated are capable of projecting plasma into a desired area (see, for example, E. Harris, R. Theus, and W. H. Bostick, Phy. Rev., 105, 46, 1957).

It should be further understood that the sources for providing the necessary discharge and vaporizing currents may be any source or group of sources capable of yielding the high electrical currents and rapid discharge time required to produce the operation described.

For many space flight missions, such as satellite orbit transfer or long space flights in the absence of gravitational forces, accelerations of about $10^{-4}g$ (where g is the acceleration of gravity) are sufficient. For vehicles of approximately 10 tons, a thrust of about 2 pounds (10^6 dynes) would be required. Such a thrust could be obtained with the three axis propulsion device shown in Figure 2 by using a bank that yielded 1,000 firings per second. Thus, if a thousand units, such as illustrated in Figure 2, are provided and each unit is fired once per second, the thrust developed would be sufficient to propel vehicles of approximately 10 tons. Considering the dimension of each unit, the thrust obtained per unit area at a firing rate of one per second would be about 50 dynes/cm.^{2}.

If fine wire were used as the primary propellant it would be necessary to re-introduce the wire after each firing. Techniques for accomplishing this could be provided by a variety of techniques which will be evident to those skilled in the art. Or, a small button plasma gun could be provided for each device, and electrode or other material used as the propellant each cycle. Operation and firing would therefore continue to provide propulsion as long as sufficient propellant material remained.

It should thus be apparent that the embodiments shown are only exemfiplary and that various modifications can be made in construction and arrangement within the scope of the present invention as defined in the appended claims.

What is claimed is:

1. A plasma propulsion system adapted for operation substantially in a vacuum environment, said system comprising in combination: first and second oppositely disposed electrodes, means applying an electrical voltage between said electrodes, a metal member serving as a "backstrap" adjacent the path between said electrodes and disposed on the side of said electrodes opposite to the direction in which thrust is desired, means for introducing a plasma between said electrodes, said propulsion system being so constructed and arranged that the introduction of said plasma between said electrodes causes the electrical voltage between said electrodes to break down the gap therebetween, said "backstrap" being connected so that a substantial portion of the discharge current through said gap flows through said "backstrap," the forces acting on said plasma upon breakdown of said gap thereby causing acceleration of said plasma away from the vicinity of said electrodes in a direction substantially away from said "backstrap."

2. A plasma propulsion system adapted for operation substantially in a vacuum environment, said system comprising in combination: first and second oppositely disposed electrodes, means applying an electrical voltage between said electrodes, a metal member serving as a "backstrap" adjacent the path between said electrodes and disposed on the side of said electrodes opposite to the direction in which thrust is desired, third and fourth oppositely disposed electrodes substantially at right angles to said first and second electrodes, a fine wire connected between said third and fourth electrodes and passing through an area in the vicinity of said first and second

electrodes, means for applying an electrical voltage between said third and fourth electrodes which when applied will vaporize and heat said wire to the plasma state, said propulsion system being so constructed and arranged that the vaporization and heating of said wire to the plasma state causes the electrical voltage between said first and second electrodes to break down the gap therebetween, said "backstrap" being connected in series with said first and second electrodes and the applied voltage thereto so that the discharge current through said gap flows through said "backstrap," the forces acting on said plasma upon breakdown of said gap thereby causing acceleration of said plasma away from the vicinity of said first and second electrodes in a direction substantially away from said "backstrap."

3. A plasma propulsion system adapted for operation substantially in a vacuum environment, said system comprising in combination: first and second oppositely disposed electrodes, a fine wire connected between said electrodes, a metal member serving as a "backstrap" adjacent the path between said electrodes and disposed on the side of said electrodes opposite to the direction in which thrust is desired, means for applying an electrical voltage between said electrodes which when applied will vaporize and heat said wire to the plasma state and initiate a discharge between said electrodes, said "backstrap" being connected in series with said electrodes and said electrical voltage source so that the discharge current through said gap flows through said "backstrap," the forces acting on said plasma when said electrical voltage is applied thereby causing acceleration of said plasma away from the vicinity of said first and second electrodes in a direction substantially away from said "backstrap."

4. A plasma propulsion system adapted for operation substantially in a vacuum environment, said system comprising in combination: first and second oppositely disposed electrodes, means applying an electrical voltage between said electrodes, means for introducing a plasma between said electrodes, said propulsion system being so constructed and arranged that the introduction of said plasma between said electrode causes the electrical voltage between said electrodes to break down the gap therebetween, and means connected in said system so that a substantial portion of the discharge current through said gap flows in a path adjacent the path between said electrodes and on the side of said electrodes opposite to the direction in which thrust is desired, the forces acting on said plasma upon breakdown of said gap thereby causing acceleration of said plasma away from the vicinity of said electrodes.

No references cited.

WHEN RESEARCHING UFOS, THERE IS A LOT OF BS TO SORT THROUGH.

FOR EXAMPLE, THIS IS NOT A UFO OVER DOWNTOWN SEATTLE. IT IS THE MOON.

THIS IS NOT A UFO ON MARS' SURFACE. IT IS GLASS ART IN THE SIDEWALK.

SOMEONE TRIED TO REPORT THIS AS A UFO THAT TOUCHED DOWN AND STARTED SPRAYING MANNA AND THE WATERS OF LIFE ON PEOPLE.

OBSERVE THE RAINBOW AS PROOF OF THE POSITIVE LIFE GIVING ENERGY.

THIS IS ACTUALLY A PICTURE OF THE WATER FEATURE AT SEATTLE CENTER ON A NICE SUMMER DAY.

IT REALLY PISSES ME OFF THAT-- <OOF!>

DO NOT LISTEN TO HIM.
YOU CAN'T HAVE A UFO WITHOUT ELEMENT 115, WHICH DOES NOT OCCUR ON EARTH.

WOULD WE LIE TO YOU? WE FLY HALF WAY ACROSS THE GALAXY TO GIVE YOU FREE RECTAL EXAMS. WE'RE THE GOOD GUYS!

GET OUT OF HERE! WE DON'T WANT TO HEAR THIS NONSENSE!

OK. WHERE'D WE LEAVE OFF? NO BS. THE FACT IS THAT UFO PROPULSION IS REALLY EASY TO UNDERSTAND.

HUMANS AREN'T SPIRITUALLY ADVANCED ENOUGH FOR THIS TECHNOLOGY!
YOU JUST NEED--

WHO ARE YOU?
I AM THE PLEIADEAN DELEGATE.

CITIZENS OF EARTH! WE, THE PLEIADEAN DELEGATION, MUST INFORM YOU THAT BECAUSE OF YOUR LACK OF SPIRITUAL DEVELOPMENT, YOU ARE DENIED ACCESS TO UFO TECHNOLOGY. ONLY WHEN YOUR HEARTS ARE PURE WILL YOU BE ABLE TO ACCESS THE STARS. WE CAN'T ALLOW YOU TO THREATEN THE HEAVENS WITH--
WHO DRESSED YOU? BILLY MEYER OR HUGH HEFNER?

HOW DARE YOU!
REALLY?

YOU ARE A MAN OF LOW SPIRITUAL DEVELOPMENT.
I'M NOT THE ONE WHO'S DRESSED LIKE LIKE A PLAYBOY BUNNY. OR A DEAN MARTIN SHOWGIRL.

LOOK, THIS IS ABOUT TECHNOLOGY, NOT ABOUT BEING SMUGLY SUPERIOR BECAUSE OF RELIGION OR THEOSOPHY.
GO HANG OUT WITH SHORTY.

OK, NOW: THE ELECTROGRAVITIC EFFECT IS EASY TO UNDERSTAND.

THE ELECTROGRAVITIC EFFECT

PROPULSION IN ONE DIRECTION BY HIGH POWERED STIMULUS. FOR A DETAILED SCIENTIFIC ANALYSIS OF SUCH IMPULSE TECHNOLOGY, PLEASE SEE PAGES 2 THROUGH 8 OF VOLUME II OF THE UFO HOW-TO SERIES, "ELECTROGRAVITICS"

HIGH VOLTAGE IMPULSES IN A CAPACITOR WILL CAUSE IT TO LURCH IN THE DIRECTION OF THE CAPACITOR'S NORTH POLE.

HIGH SPEED ROTATION OF ELEMENTS WITH ODD NUCLEAR SPIN VALUES CAUSE DIRECTIONAL MOVEMENT THAT IS NEITHER CENTRIFUGAL NOR CENTRIPETAL MOTION.

MATERIALS WITH ODD NUCLEAR SPIN VALUES INCREASE THE PROPULSIVE EFFECT REGARDLESS OF WHETHER THE HIGH SPEED ROTATION METHOD OR THE HIGH VOLTAGE ELECTRICAL IMPULSE METHOD IS USED.

THE SECRET OF ODD NUCLEAR SPIN VALUES REGARDING THE ELECTROGRAVITIC EFFECT WAS REVEALED BY THE WORK OF HENRY W. WALLACE. FOR MORE DETAILED INFORMATION ON THE BRILLIANT WORK OF

WALLACE , PLEASE SEE VOLUME II OF THE UFO HOW-TO SERIES, "ELECTROGRAVITICS" PAGES 80 THROUGH 103.

ODD NUCLEAR SPIN VALUES ARE THE KEY TO ELECTROGRAVITICS.

BARIUM TITANATE IS PERFECT FOR T.T. BROWN'S ELECTROKINETICS BECAUSE OF A HIGH "K" DIELECTRIC AND THE DIFFERENTIAL OF NUCLEAR SPIN VALUES IN THE MATERIAL.

BARIUM +3/2 TITANIUM -7/2

ODD NUCLEAR SPIN VALUE MATERIALS CAN BE SEEN IN JOHN SEARL'S LEVITY DISK RUNNERS, WHERE THIS DIFFERENTIAL AND ITS EFFECTIVENESS ARE TESTIFIED TO BY THE RUNNERS' COMPOSITION:

ALUMINUM +5/2 NEODYMIUM -7/2

A SIGNIFICANT DIFFERENTIAL FOR THE ELECTROGRAVITIC EFFECT!

MUMETAL, SPOKEN OF BY PROFESSOR FRAN DEAQUINO IN VOLUME II OF THE UFO HOW-TO SERIES, ALSO FOLLOWS IN THIS REGARDS, COMPOSED OF:

NICKEL -3/2 NSV
MOLYBDENUM +5/2 NSV OR -5/2 NSV

LINDA MOULTON HOWE'S "MYSTERY METAL FRAGMENT" IS COMPOSED OF:

BISMUTH: +9/2 NSV
MAGNESIUM: -5/2 NSV
WITH 3% ZINC: +5/2 NSV
(NSV = NUCLEAR SPIN VALUE)

WHERE THE EVIDENCE CAN BE FOUND, IT SEEMS TO POINT TO MATERIALS WITH A HIGH DIFFERENTIAL IN THE ODD NUCLEAR SPIN VALUE MATERIALS HAVING THE GREATEST ABILITY TO MAXIMIZE THE ELECTROGRAVITIC EFFECT:

COMPARISONS:

BARIUM TITANATE	
BARIUM	+3/2 NSV
TITANIUM	-7/2 NSV

JOHN SEARL'S LEVITY DISK	
ALUMINUM	+5/2 NSV
NEODYMIUM	-7/2 NSV

LINDA HOWE'S "MYSTERY METAL FRAGMENT"	
BISMUTH:	+9/2 NSV
MAGNESIUM:	-5/2 NSV
WITH 3% ZINC:	+5/2 NSV

MUMETAL	
NICKEL	-3/2 NSV
MOLYBDENUM	+5/2 NSV OR -5/2 NSV

BISMANOL	
BISMUTH	+9/2 NSV
MANGANESE	+5/2 NSV

BUT WHAT EXACTLY IS THE ELECTROGRAVITIC EFFECT? HOW CAN WE VISUALIZE IT IN OUR MIND'S EYE? ONE WAY IS AS FOLLOWS:

A HIGH VOLTAGE CURRENT

IS PULSED INTO MATERIALS WITH ODD NUCLEAR SPIN VALUES.

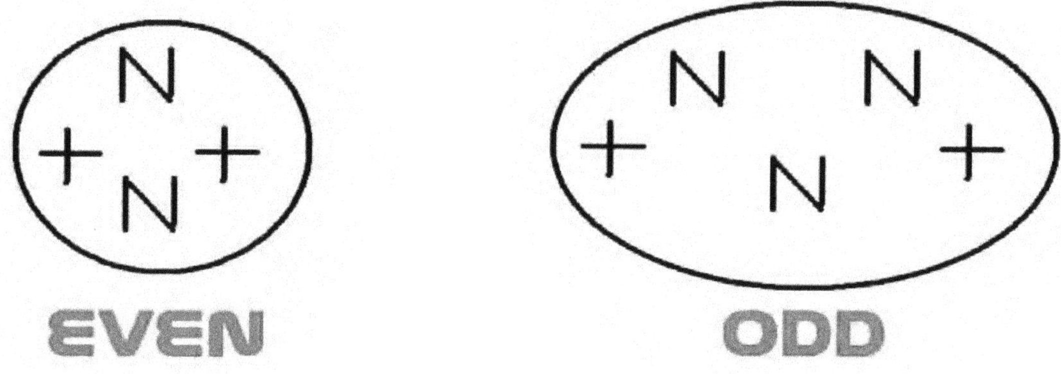

THE MATERIALS WITH ODD NUCLEAR SPIN VALUES CAN BE CONCEIVED OF AS DEFORMING INTO AN EGG-SHAPE IN THE DIRECTION OF THEIR COLLECTIVE NORTH POLES UNDER STIMULATION.

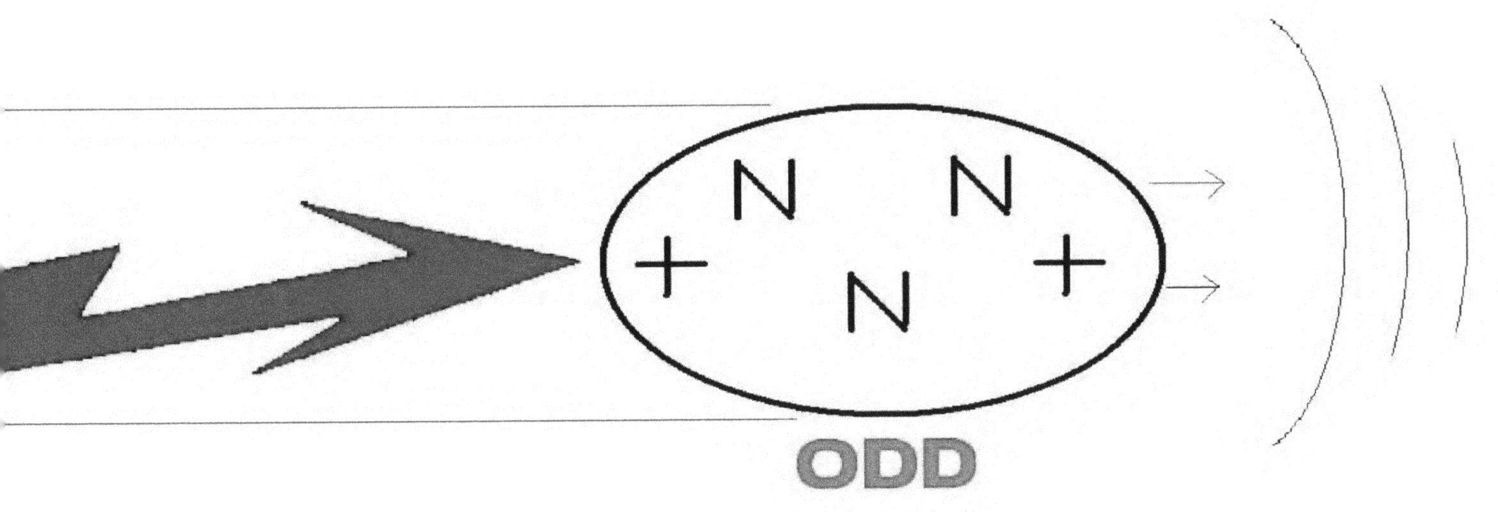

THE OVERALL STRUCTURE OF THE UCLEUS WAS WEAKER THAN THE ΛPULSES BEING IMPRESSED, ATASTROPHIC FAILURE COULD OCCUR. WITH /2 AND 7/2 MATERIALS, EITHER POSITIVE R NEGATIVE, THIS DOES NOT APPEAR TO BE CONCERN.

OWEVER, MATERIALS WITH 9/2 ODD UCLEAR SPIN VALUES ARE USUALLY ADIOACTIVE.

SO WHILE A BISMUTH/STRONTIUM COMBINATION OR BISMUTH/URANIUM MIGHT SEEM TO BE IDEAL WITH +9/2 AND -9/2 ODD NUCLEAR SPIN VALUE DIFFERENTIALS, THE THREAT AND DANGER OF THE RADIOACTIVE MATERIALS IS TOO GREAT A BARRIER TO OVERCOME.

SO, GENERALLY SPEAKING:

THE HIGHER THE APPLIED VOLTAGE, THE GREATER THE NUMBER / FREQUENCY OF PULSES, THE GREATER THE DIFFERENTIAL OF ODD NUCLEAR SPIN VALUES IN THE RECEIVING MATERIALS, THE MORE POWERFUL THE ELECTROGRAVITIC EFFECT APPEARS TO OCCUR.

THE HIGHER THE VOLTAGE,
+
THE HIGHER FREQUENCY OF PULSES,
+
THE GREATER THE DIFFERENTIAL OF ODD NUCLEAR SPIN VALUES,

=

THE MORE POWERFUL ELECTROGRAVITIC EFFECT.

YOU CAN LEARN MORE ABOUT ELECTROGRAVITICS AND THE RELATED SCIENCE IN VOLUME II OF THE UFO HOW-TO SERIES: "ELECTROGRAVITICS."

ELEMENTS AND THEIR NUCLEAR SPIN VALUES

Atom/Isotope	Spin
Hydrogen 1H	+1/2
Deuterium 2D	1
Helium 3He	-1/2 or 0
Lithium 6Li	+1
Lithium 7Li	+3/2
Beryllium 9Be	-3/2
Boron 10B	+3
Boron 11B	+3/2
Carbon 13C	+1/2 or 0
Nitrogen 14N	+1
Nitrogen 15N	-1/2
Oxygen 17O	-5/2
Fluorine 19F	+1/2
Neon 21Ne	+3/2
Sodium 23Na	+3/2
Magnesium 25Mg	-5/2
Aluminum 27Al	+5/2
Silicon 29Si	-1/2
Phosphorus 31P	+1/2
Sulfur 33S	+3/2
Chlorine 35Cl	+3/2
Chlorine 37Cl	+3/2
Potassium 39K	+3/2
Potassium 41K	+3/2
Calcium 43Ca	-7/2
Scandium 45Sc	+7/2
Titanium 47Ti	-5/2
Titanium 49Ti	-7/2
Vanadium 50V	+6
Vanadium 51V	+7/2
Chromium 53Cr	-3/2
Manganese 55Mn	+5/2
Iron 57Fe	+1/2
Cobalt 59Co	+7/2
Nickel 61Ni	-3/2
Copper 63Cu	+3/2
Copper 65Cu	+3/2
Zinc 67Zn	+5/2
Gallium 69Ga	+3/2
Gallium 71Ga	+3/2
Germanium 73Ge	-9/2
Arsenic 75As	+3/2
Selenium 77Se	+1/2
Bromine 79Br	+3/2
Bromine 81Br	+3/2
Krypton 83Kr	-9/2
Rubidium 85Rb	+5/2
Rubidium 87Rb	+3/2
Strontium 87Sr	-9/2
Yttrium 89Y	-1/2
Zirconium 91Zr	-5/2
Niobium 93Nb	+9/2
Molybdenum 95Mo	+5/2
Molybdenum 97Mo	-5/2
Ruthenium 99Ru	-5/2
Ruthenium 101Ru	-5/2

Atom/Isotope	Spin
Rhodium 103Rh	-1/2
Palladium 105Pd	-5/2
Silver 107Ag	-1/2
Silver 109Ag	-1/2
Cadmium 111Cd	-1/2
Cadmium 113Cd	-1/2
Indium 113In	+9/2
Indium 115In	+9/2
Tin 117Sn	-1/2
Tin 119Sn	-1/2
Antimony 121Sb	+5/2
Antimony 123Sb	+7/2
Tellurium 123Te	-1/2
Tellurium 125Te	-1/2
Iodine 127I	+5/2
Xenon 129Xe	-1/2
Xenon 131Xe	+3/2
Cesium 133Cs	+7/2
Barium 135Ba	+3/2
Barium 137Ba	+3/2
Lanthanum 138La	+5
Lanthanum 139La	0
Hafnium 177Hf	+7/2
Hafnium 179Hf	-9/2
Tantalum 181Ta	+7/2
Tungsten 183W	+1/2
Rhenium 185Re	+5/2
Osmium 187Os	+1/2
Osmium 189Os	+3/2
Rhenium 187Re	+5/2
Iridium 191Ir	+3/2
Iridium 193Ir	+3/2
Platinum 195Pt	+1/2
Gold 197Au	+3/2
Mercury 199Hg	+1/2
Mercury 201Hg	-3/2
Thallium 203Tl	+1/2
Thallium 205Tl	+1/2
Lead 207Pb	+1/2
Bismuth 209Bi	+9/2
Praseodymium	+5/2
Neodymium	-7/2
Samarium	-7/2
Europium	+5/2
Gadolinium	-3/2
Terbium	+3/2
Dysprosium 161Dy	-5/2
Dysprosium 163Dy	+5/2
Holmium	+7/2
Erbium 167Er	-7/2
Thulium	-1/2
Ytterbium 171Yt	+1/2
Ytterbium 173Yt	-5/2
Lutetium 175Lu	+7/2
Lutetium 175.5Lu	+7
Uranium 235Ur	+7/2

United States Patent — **Patent Number: 4,663,932**
Cox — **Date of Patent: May 12, 1987**

DIPOLAR FORCE FIELD PROPULSION SYSTEM

Inventor: **James E. Cox,** 5455 Romaine St., Los Angeles, Calif. 90038

Appl. No.: **401,526**

Filed: **Jul. 26, 1982**

Int. Cl.[4] **F03H 5/00**
U.S. Cl. **60/200.1;** 60/202; 313/359.1; 315/5.41
Field of Search 60/202, 203.1, 200.1; 313/359.1, 361.1, 362.1; 315/111.01, 5.41, 5.42

References Cited

U.S. PATENT DOCUMENTS

3,095,163	6/1963	Hill	244/12
3,322,374	5/1967	King	244/62
3,324,316	6/1967	Cann	310/11
3,334,022	9/1967	Eckert	313/63
3,353,354	11/1967	Friedman et al.	60/203.1
3,371,490	3/1968	Haslund	60/202
3,505,550	4/1970	Levoy et al.	313/63
3,527,055	9/1970	Rego	60/224
3,662,554	5/1972	Broqueville	60/202
3,678,306	7/1972	Garnier et al.	310/11
3,735,591	5/1973	Burkhart	60/202
3,866,414	2/1975	Bahn	60/202

OTHER PUBLICATIONS

Jahn, R. G., *Physics of Electric Propulsion,* McGraw-Hill, 1968, pp. 196-198.
Cox, J. E., "Electromagnetic Propulsion without Ionization", ATAA/SAE/ASME 16th Joint Propulsion Conference Hartford, Conn., 1980.

Primary Examiner—Louis J. Casaregola
Attorney, Agent, or Firm—Daniel J. Meaney, Jr.

ABSTRACT

A dipolar force field propulsion system having a alternating electric field source for producing electromotive lines of force which extend in a first direction and which vary at a selected frequency and having an electric field strength of a predetermined magnitude, a source of an alternating magnetic field having magnetic lines of force which extend in a second direction which is at a predetermined angle to the first direction of the electromotive lines of force and which cross and intercept the electromotive line of force at a predetermined location defining a force field region and wherein the frequency of the alternating magnetic field substantially equal to the frequency of the alternating electric field and at a selected in phase angle therewith and wherein the magnetic field has a flux density which when multiplied times the selected frequency is less than a known characteristic field ionization potential limit; a source of neutral particles of matter having a selected dipole characteristic and having a known characteristic field ionization potential limit which is greater than the magnitude of the electric field and wherein the dipoles of the particles of matter are capable of being driven into cyclic rotation at the selected frequency by the electric field to produce a reactive thrust, a vaporizing stage which vaporizes said particles of matter into a gaseous state at a selected temperature, and a transporting system for transporting the vaporized particles of matter into the force field defined by the crossing electromotive lines of force and the magnetic lines of force.

21 Claims, 53 Drawing Figures

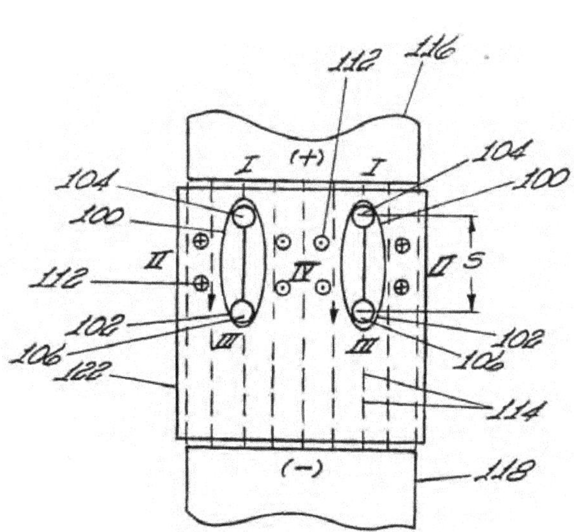

U.S. Patent May 12, 1987 Sheet 1 of 16 4,663,932

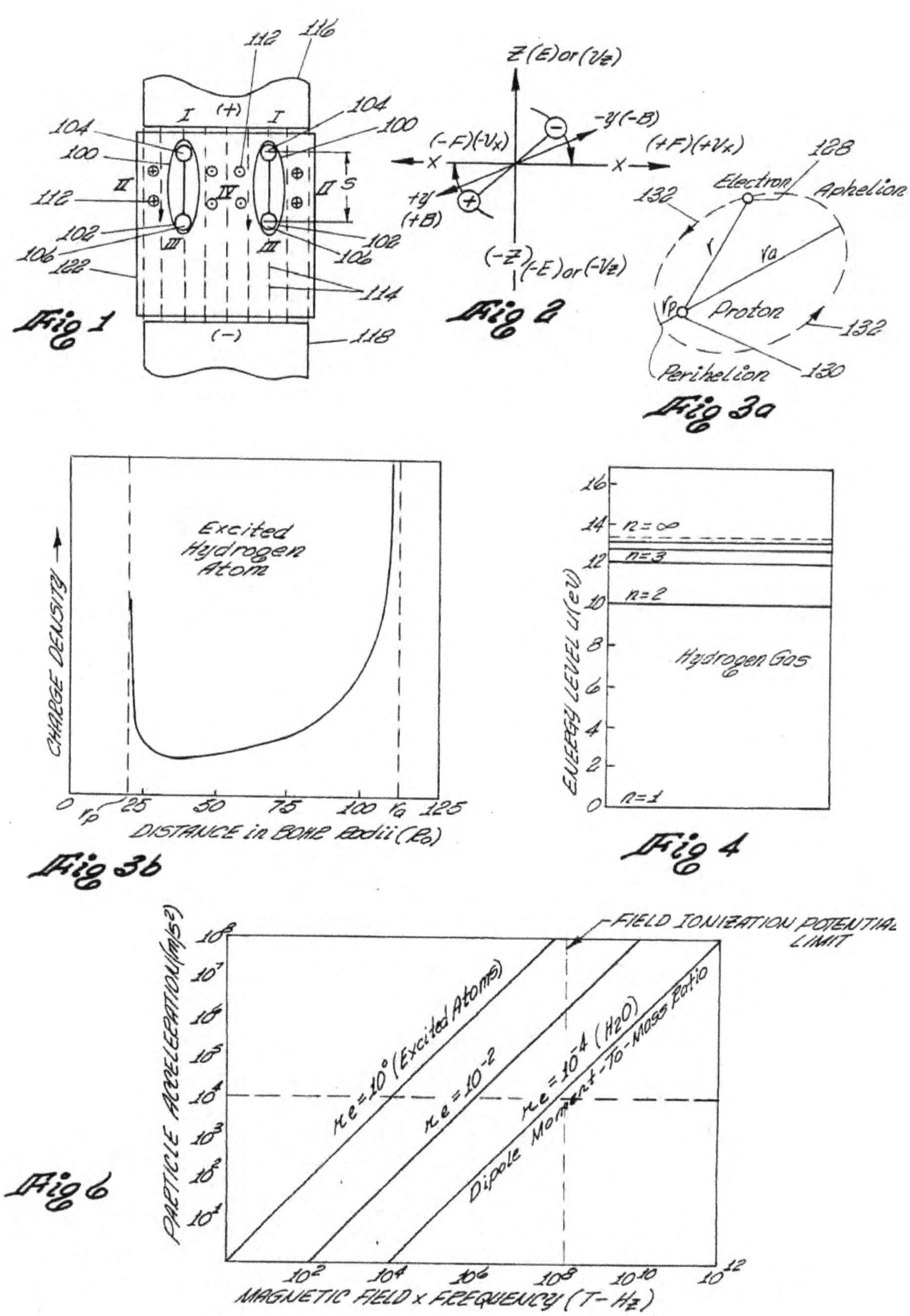

U.S. Patent May 12, 1987 Sheet 2 of 16 4,663,932

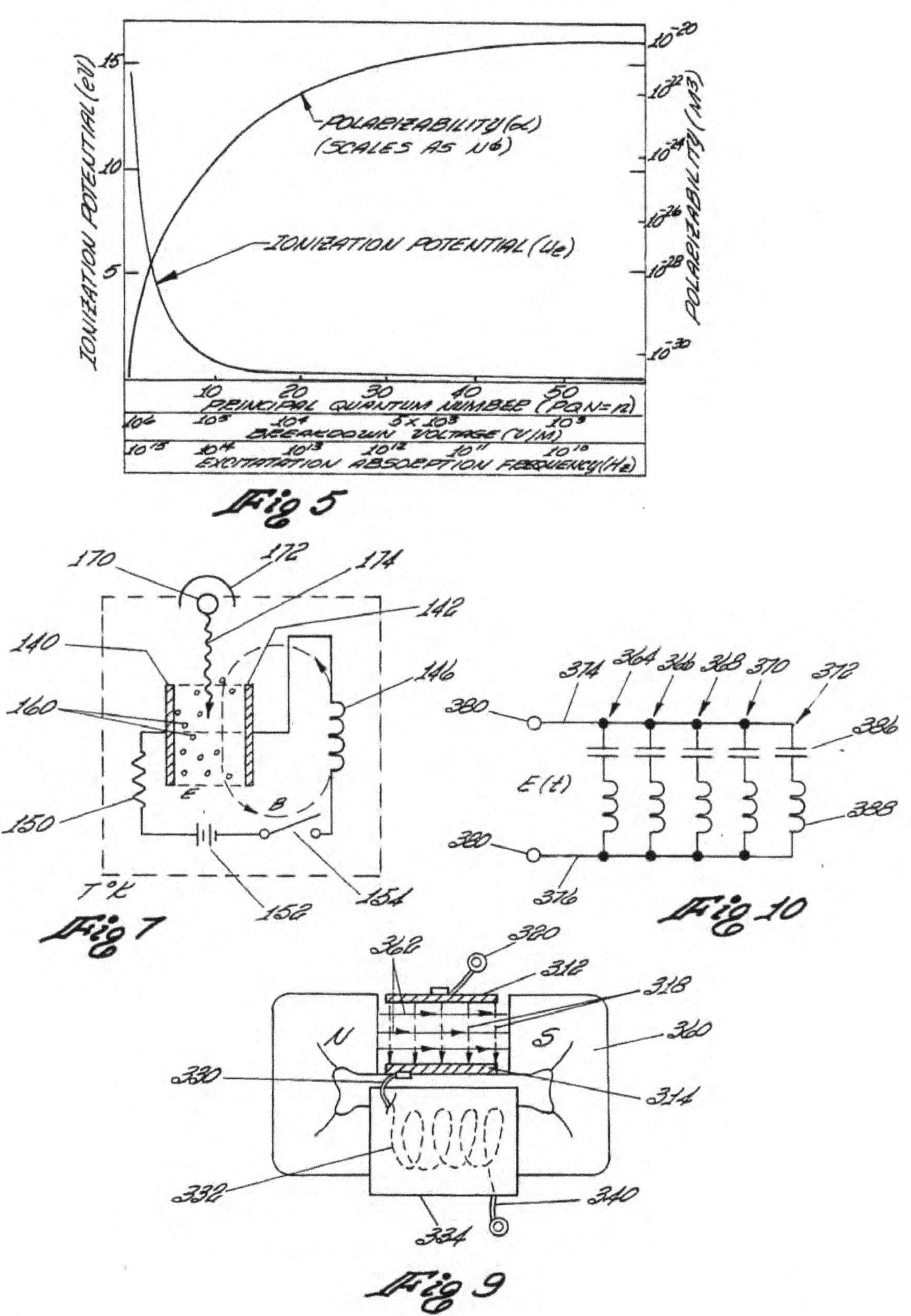

U.S. Patent May 12, 1987 Sheet 3 of 16 4,663,932

Fig 8

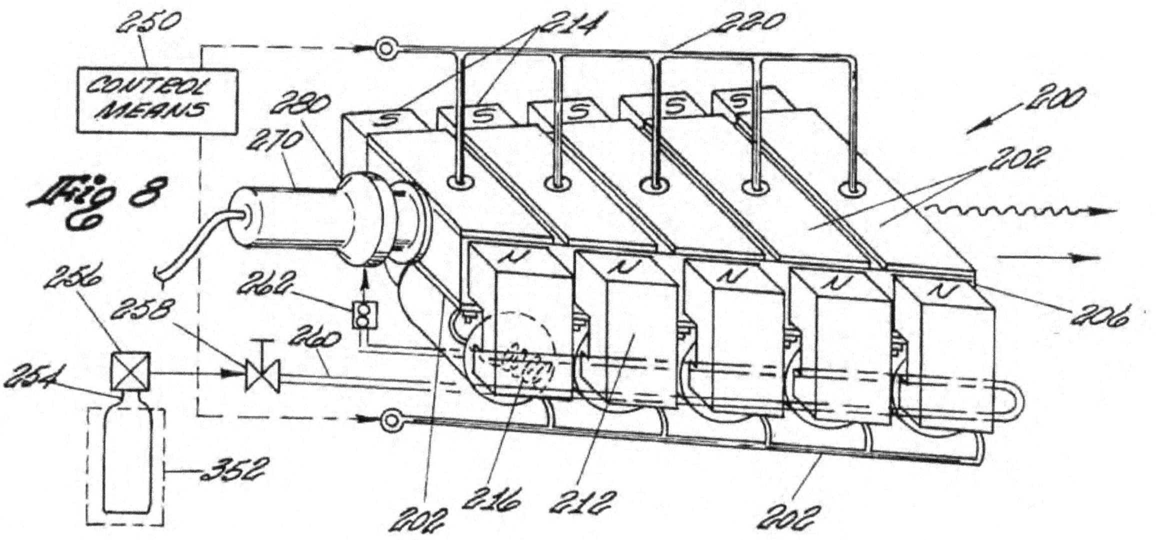

250 CONTROL MEANS 214 220 280 270 200 202 256 262 258 260 254 206 352 202 216 212 202

S S S S S

N N N N N

Fig 11

SPECIFIC IMPULSE (SECS)

10^0 10^1 10^2 10^3 10^4

1×10^0 (EXCITED GASES)

1×10^{-2}

1×10^{-4} (GROUND STATE WATER)

1×10^{-6} DIPOLE MOMENT/MASS RATIO

10^4 10^6 10^8 10^{10} 10^{12} 10^{14}

OPERATING PARAMETER (L·V·B)

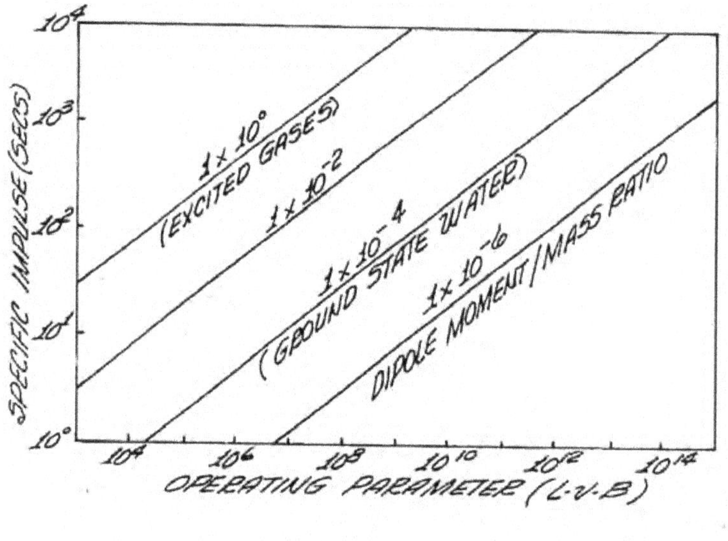

Fig 12

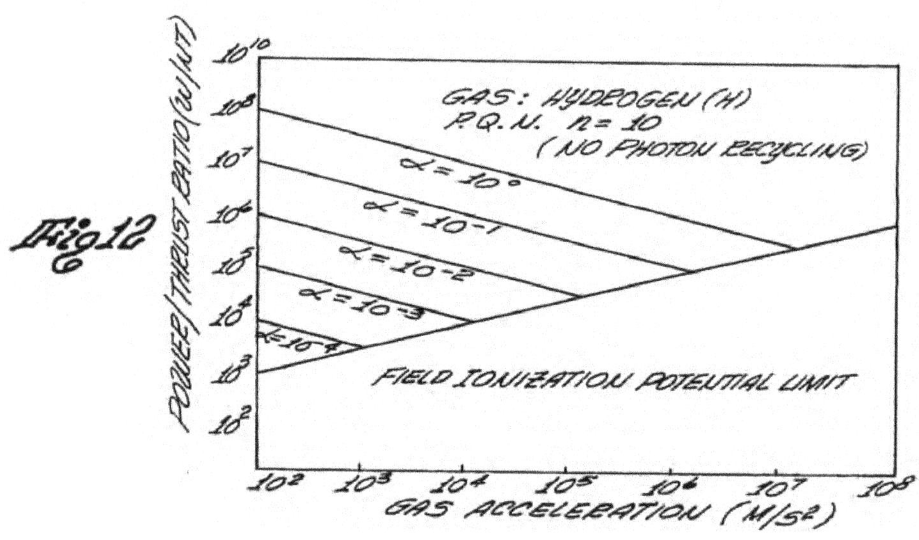

U.S. Patent May 12, 1987 Sheet 4 of 16 4,663,932

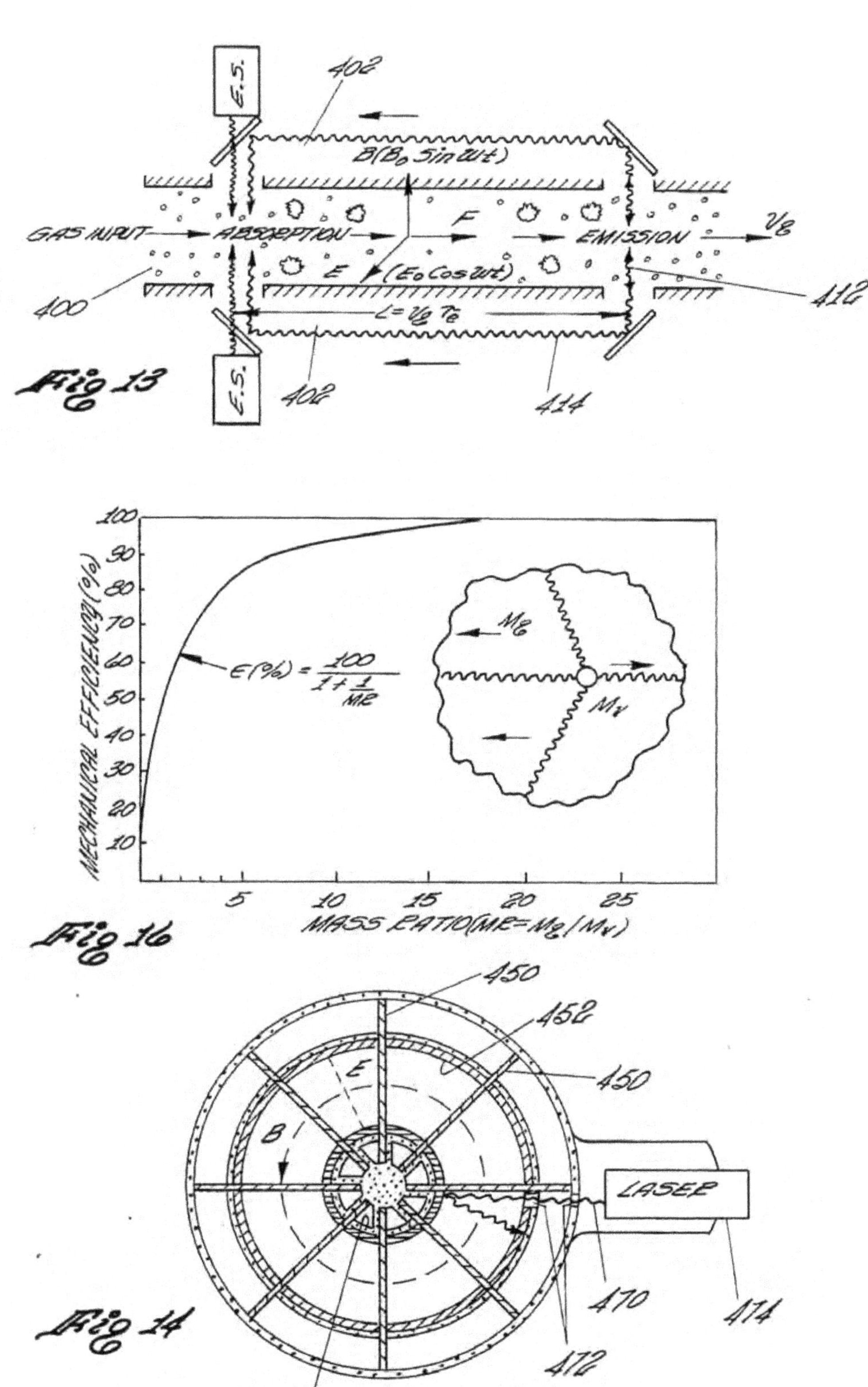

U.S. Patent May 12, 1987 Sheet 5 of 16 4,663,932

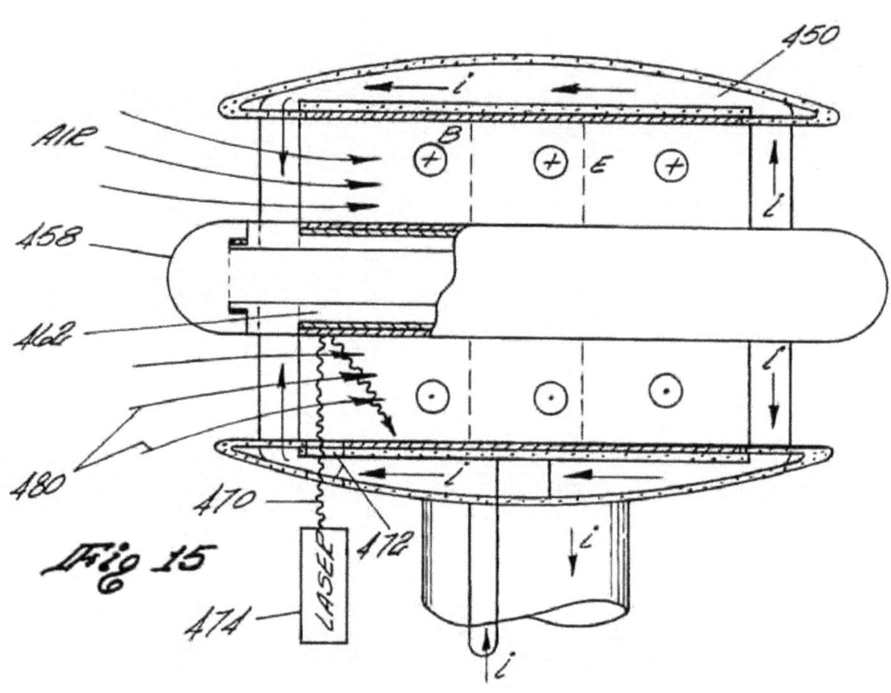

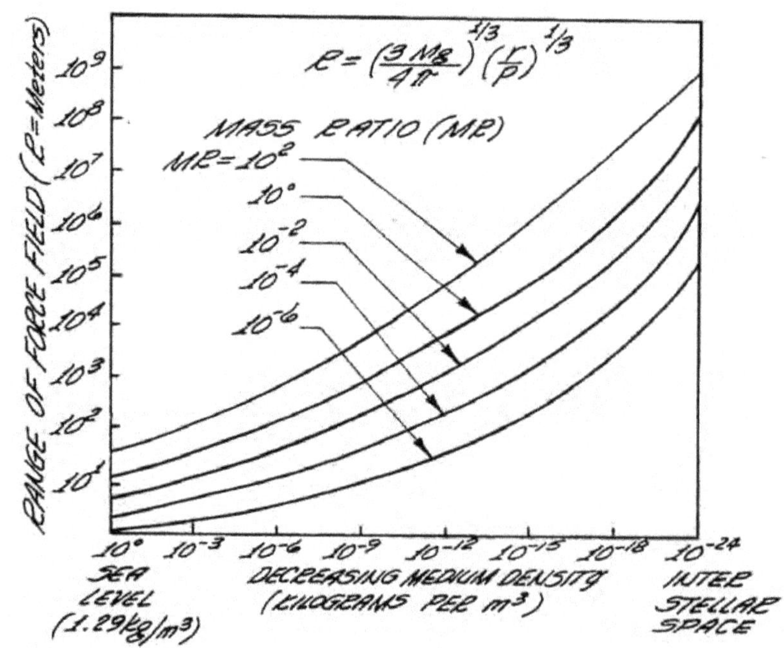

U.S. Patent May 12, 1987 Sheet 6 of 16 4,663,932

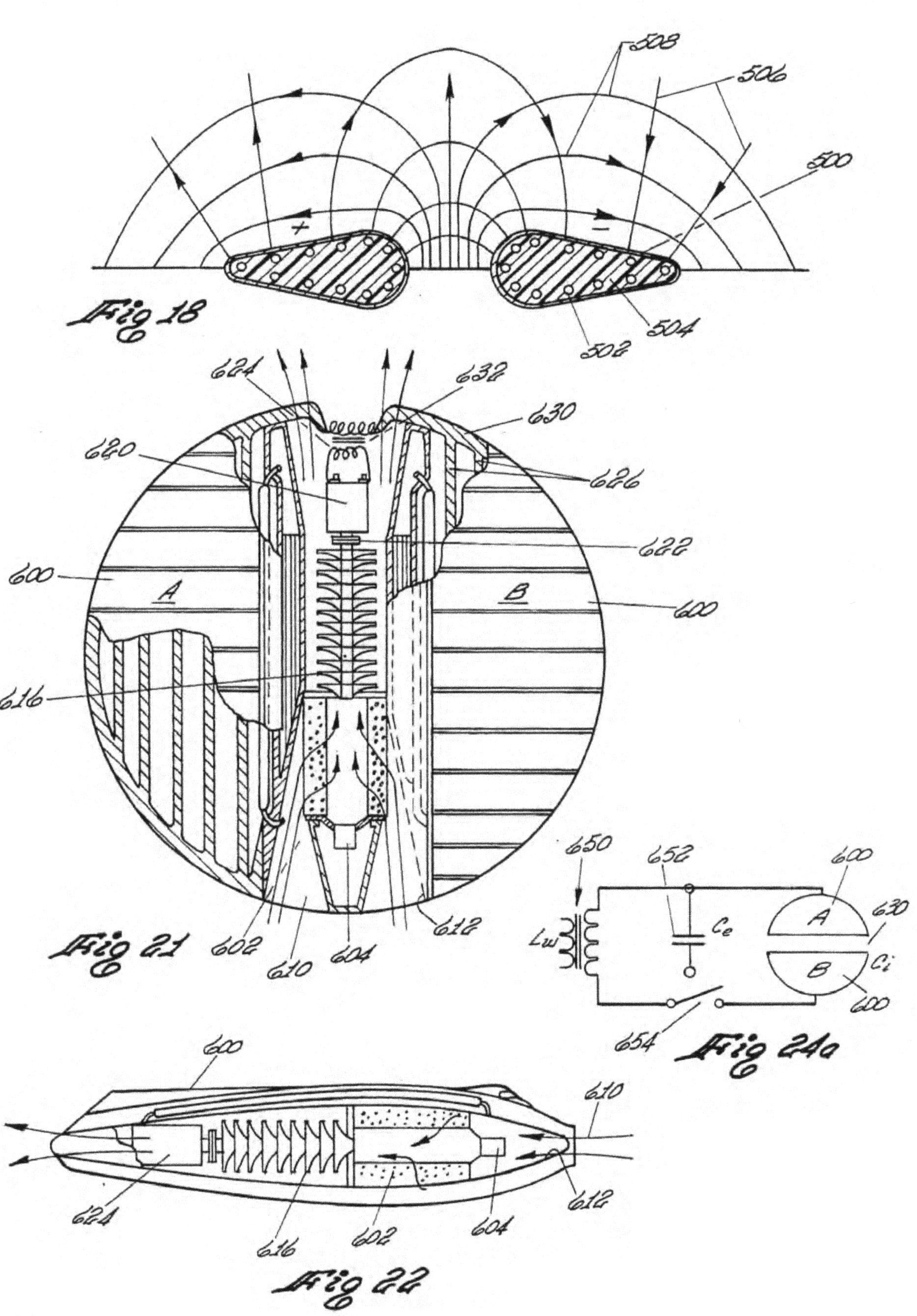

U.S. Patent May 12, 1987 Sheet 7 of 16 4,663,932

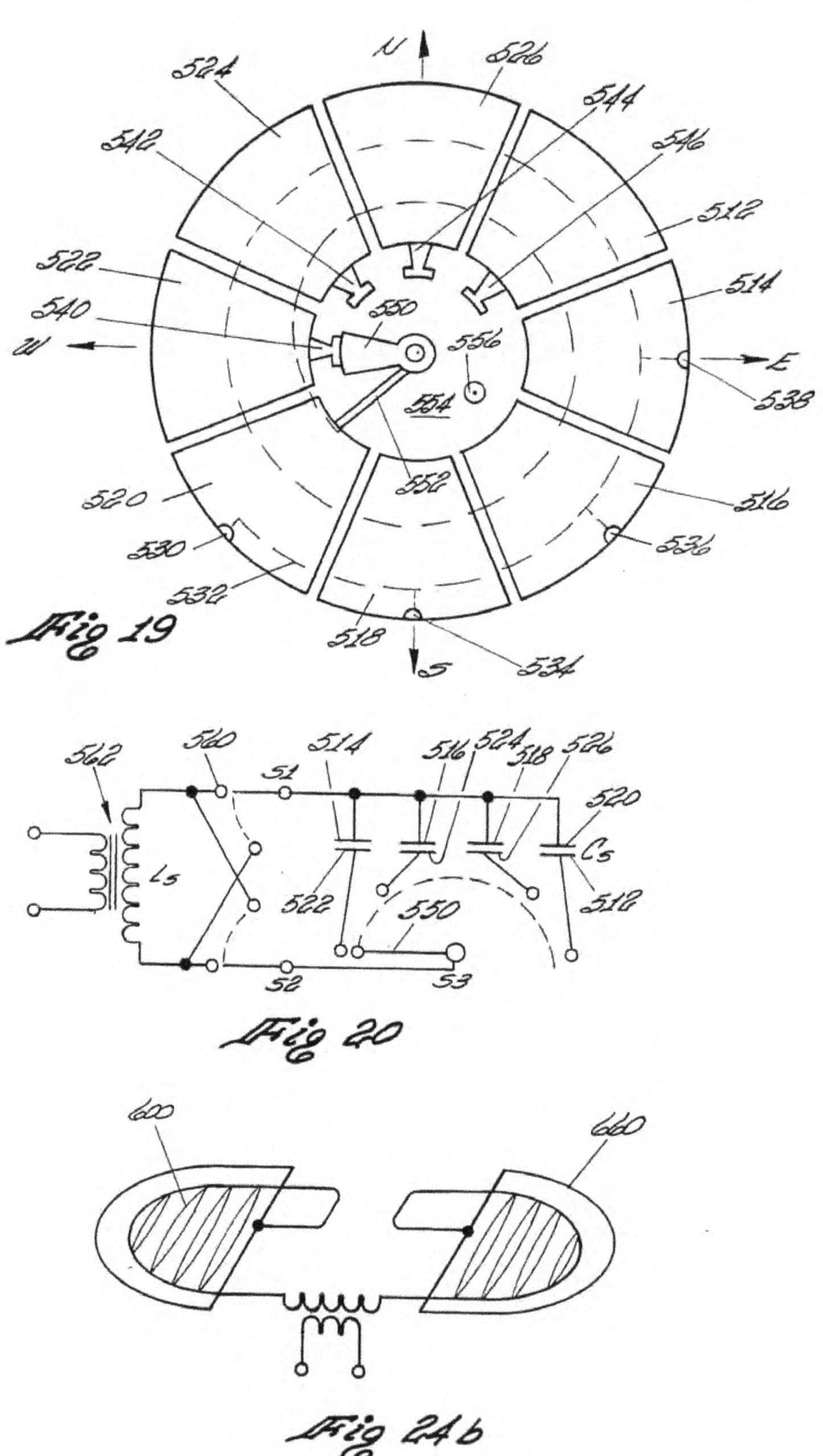

Fig 19

Fig 20

Fig 24b

U.S. Patent May 12, 1987 Sheet 8 of 16 4,663,932

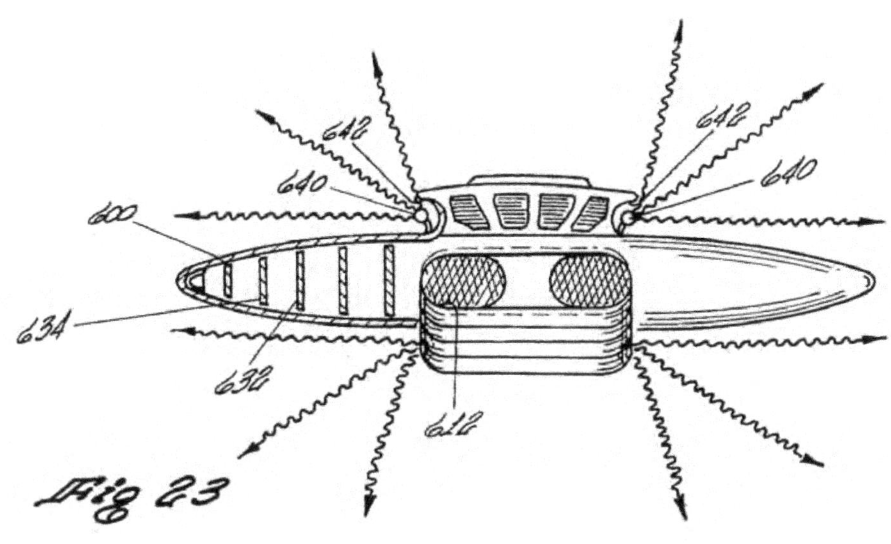

Fig 23

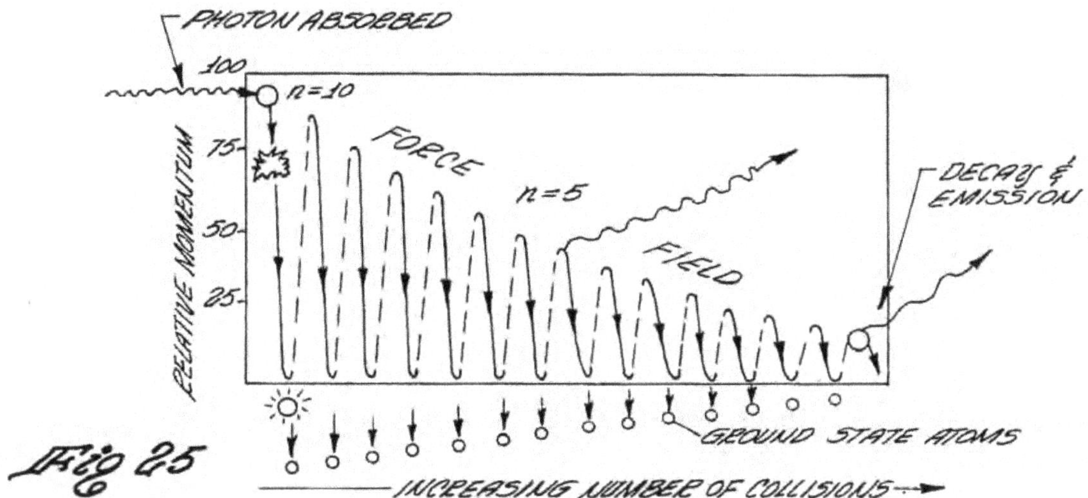

Fig 25

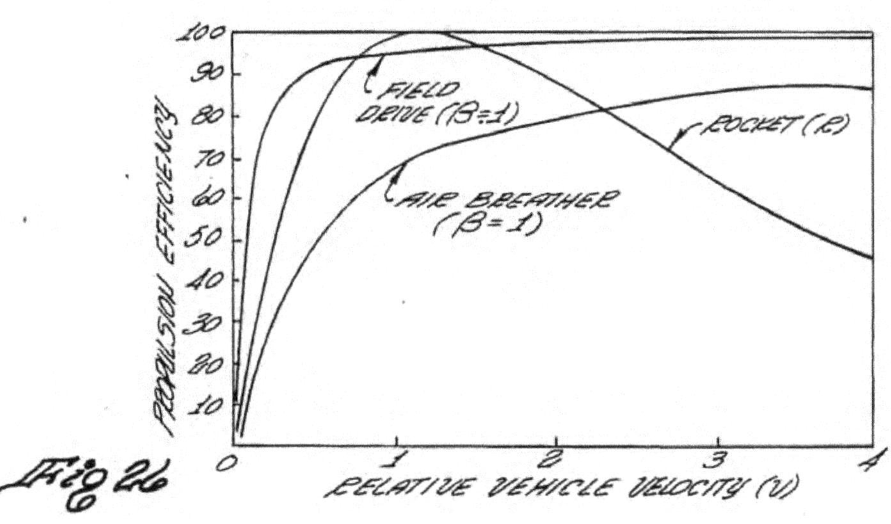

Fig 26

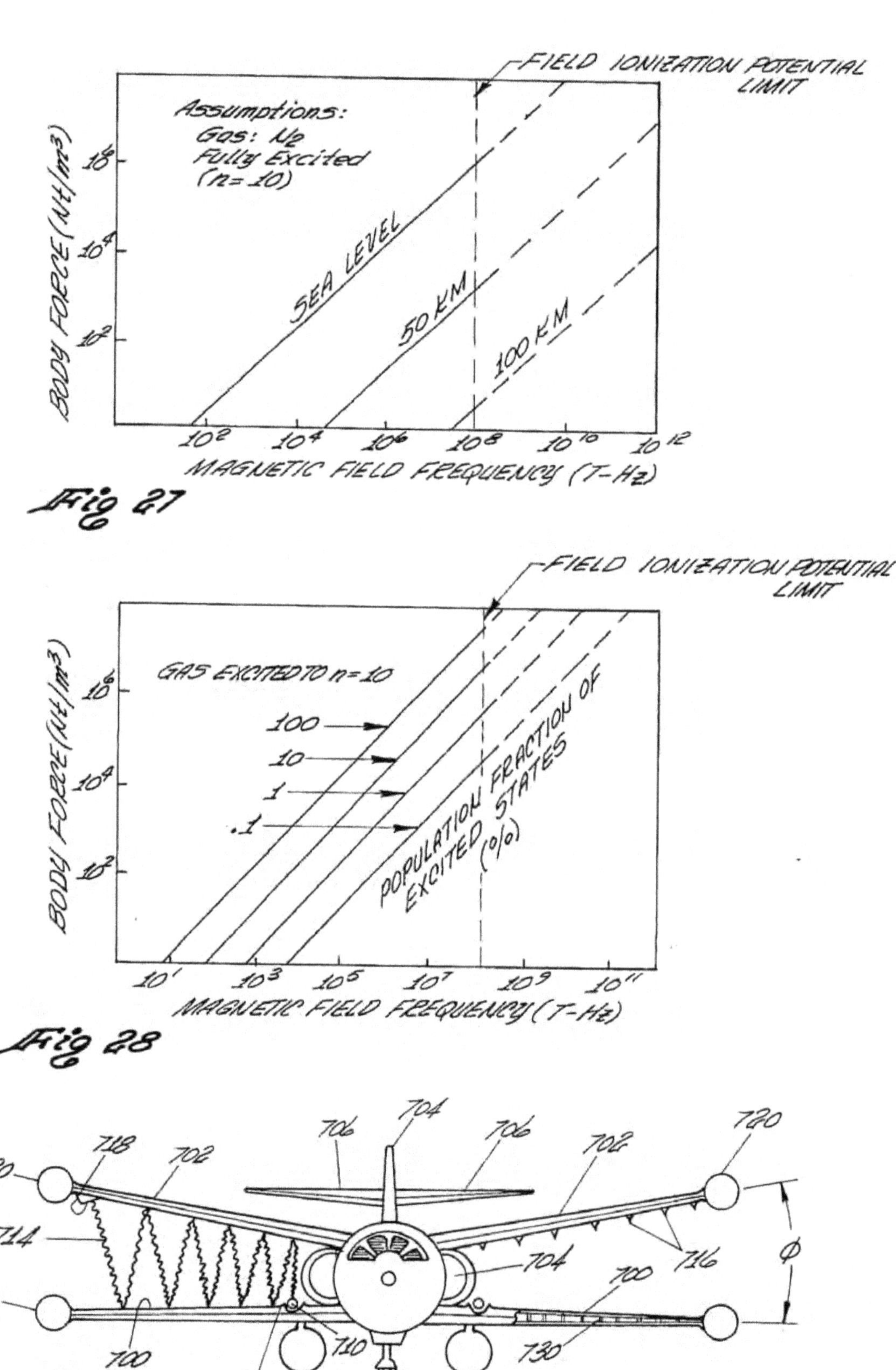
FIELD IONIZATION POTENTIAL LIMIT
Assumptions:
Gas: N2
Fully Excited
(n= 10)
SEA LEVEL
50 KM
100 KM
BODY FORCE (Nt/m3)
MAGNETIC FIELD FREQUENCY (T-Hz)
Fig 27
FIELD IONIZATION POTENTIAL LIMIT
GAS EXCITED TO n=10
100
10
1
.1
POPULATION FRACTION OF EXCITED STATES (%)
BODY FORCE (Nt/m3)
MAGNETIC FIELD FREQUENCY (T-Hz)
Fig 28
704
706
706
718
702
702
720
720
720
720
714
704
700
716
700
710
708
730
726
724
Fig 29

U.S. Patent May 12, 1987 Sheet 10 of 16 4,663,932

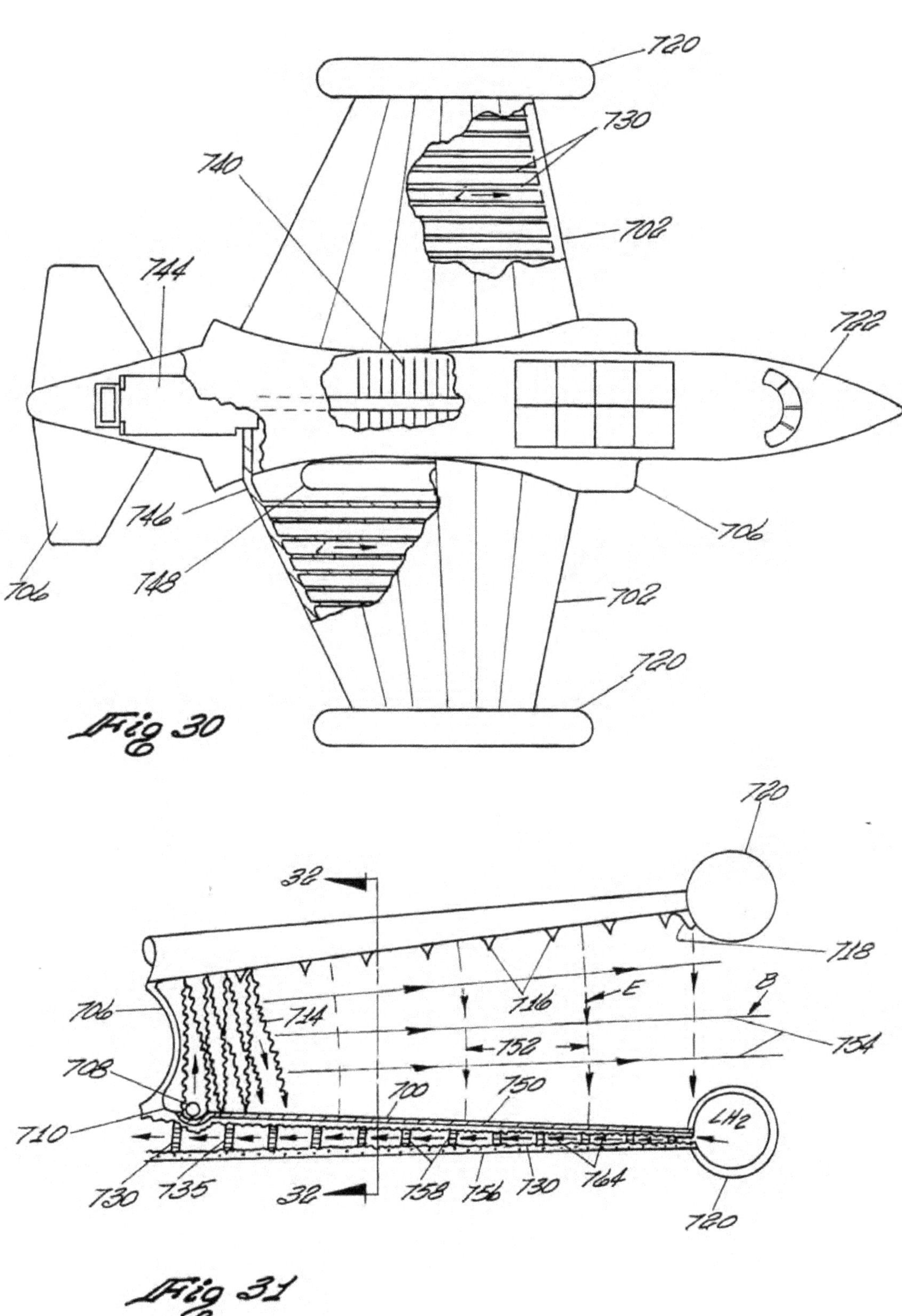

Fig 30

Fig 31

U.S. Patent May 12, 1987 Sheet 11 of 16 4,663,932

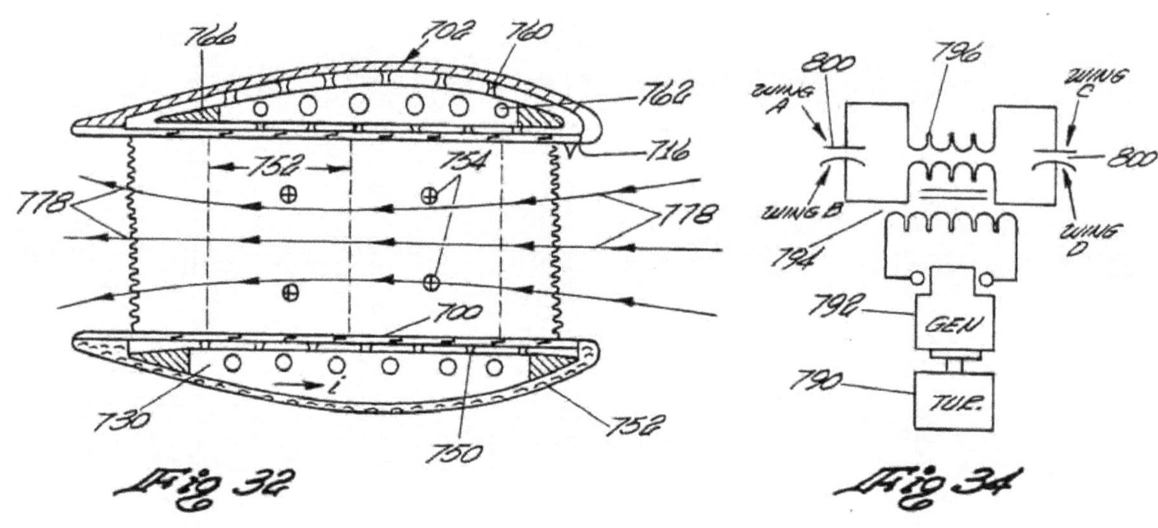

Fig 32

Fig 34

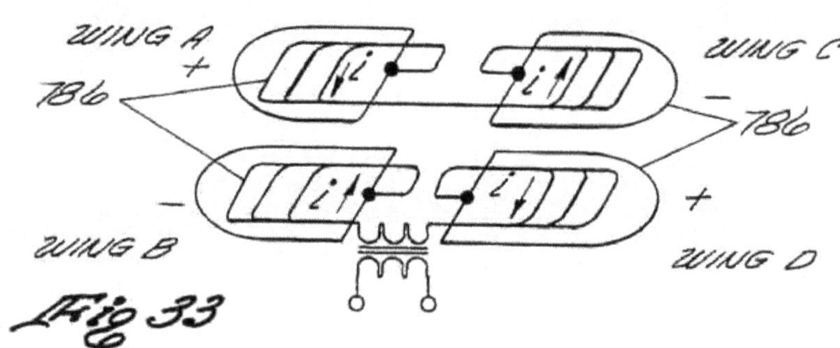

Fig 33

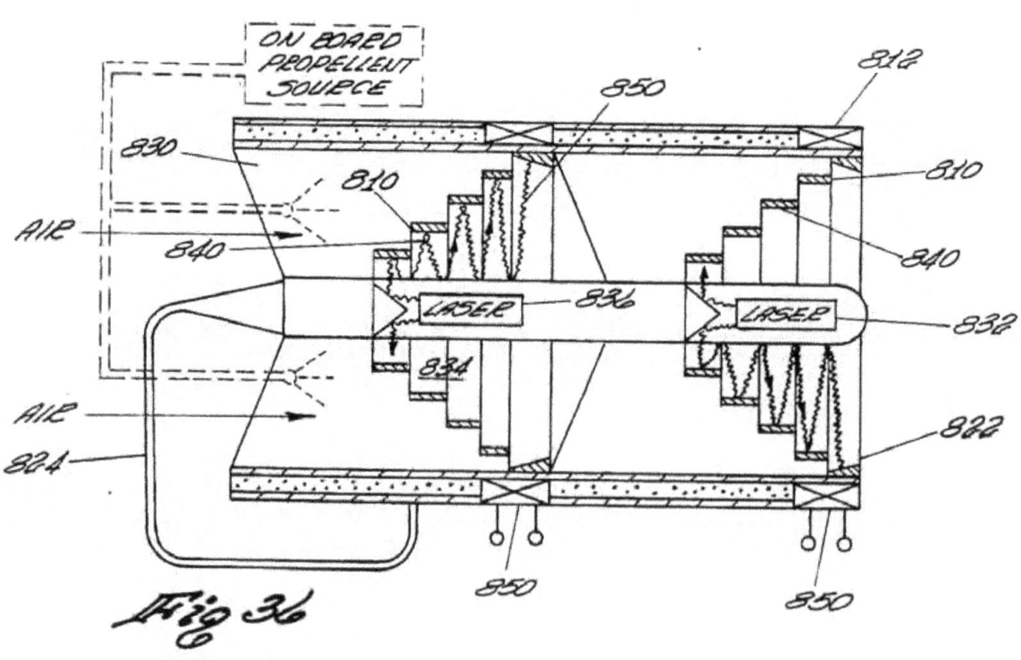

Fig 36

U.S. Patent May 12, 1987 Sheet 12 of 16 4,663,932

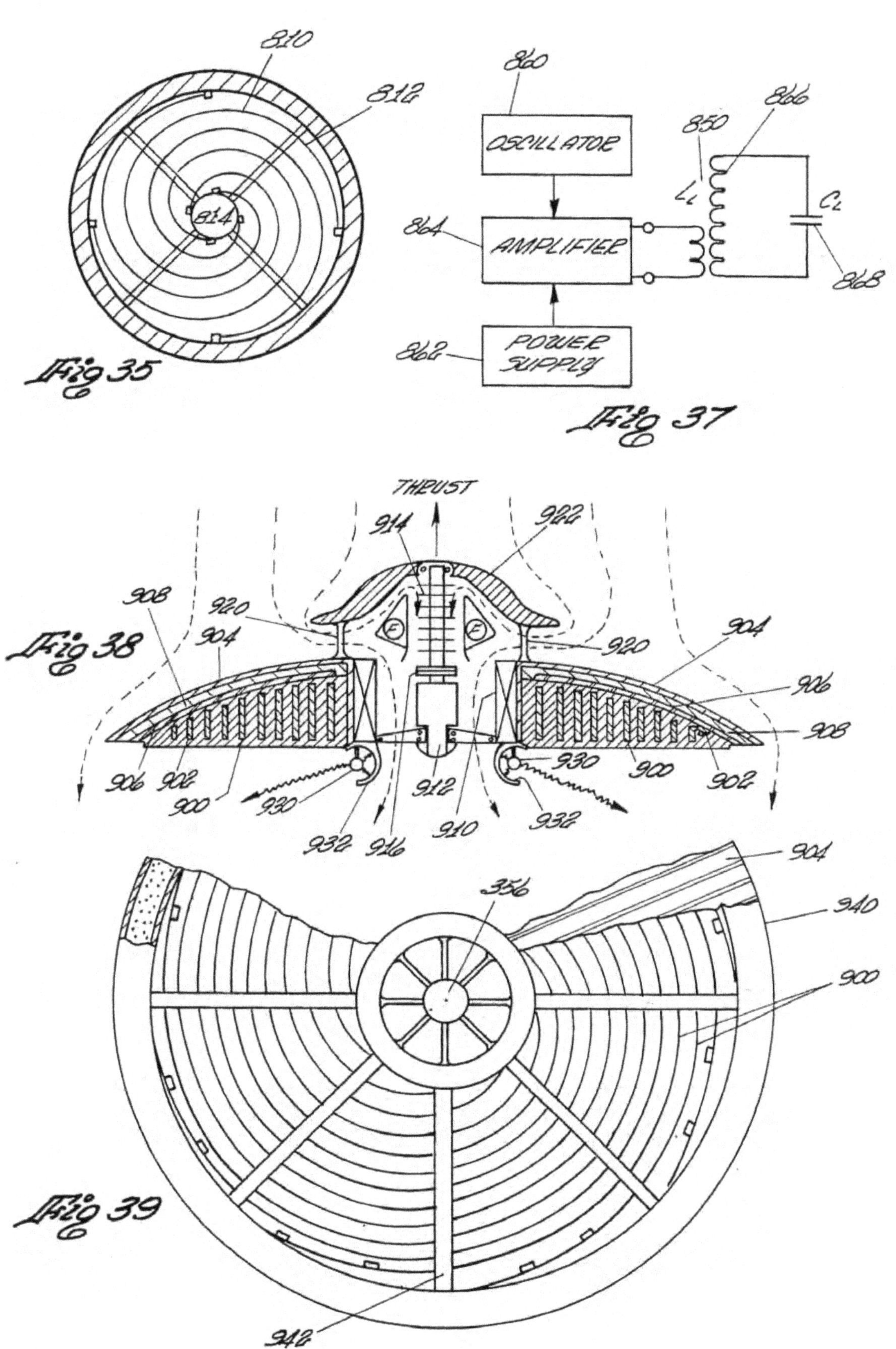

U.S. Patent May 12, 1987 Sheet 13 of 16 **4,663,932**

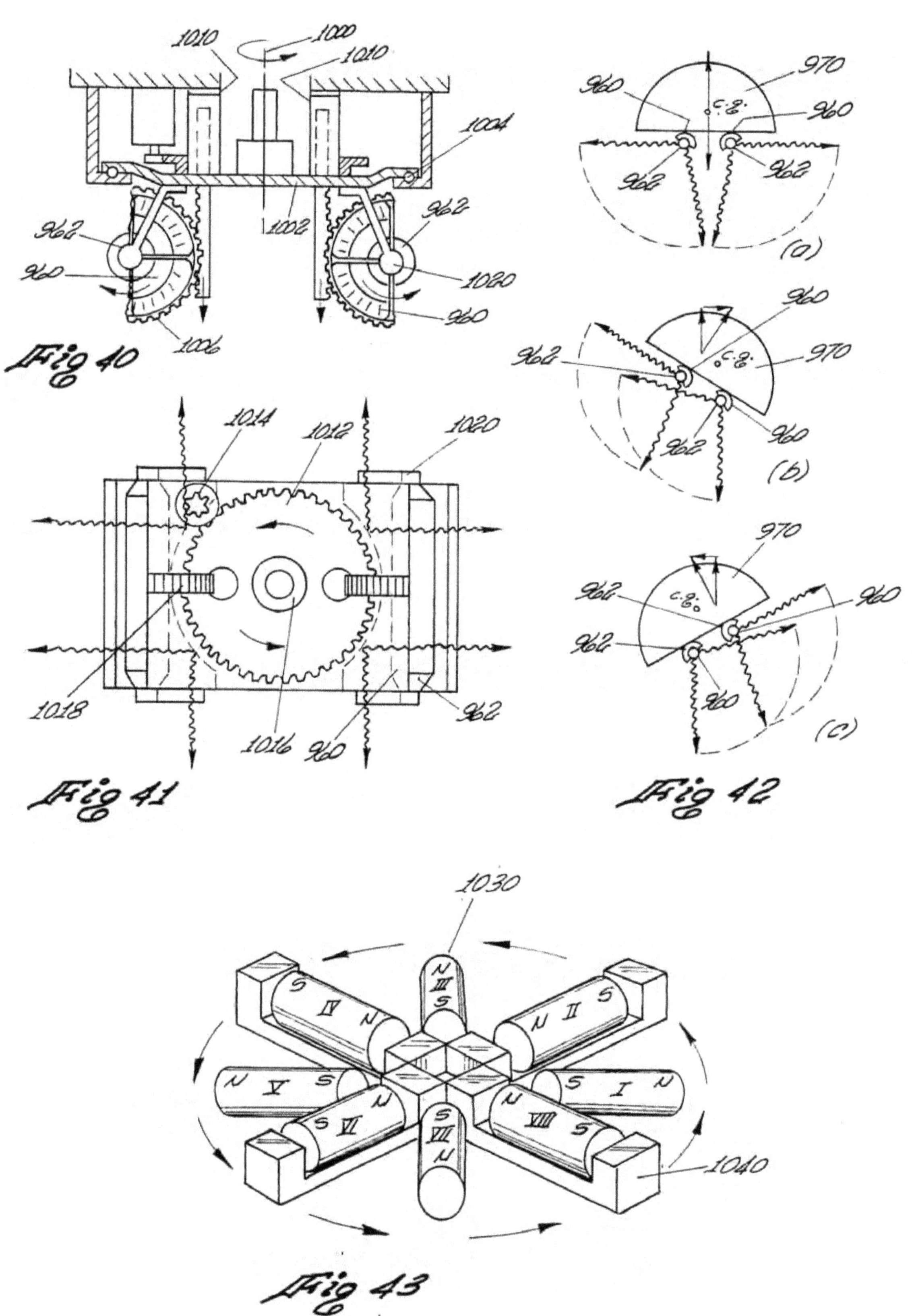

Fig 40

Fig 41

Fig 42

Fig 43

U.S. Patent May 12, 1987 Sheet 14 of 16 4,663,932

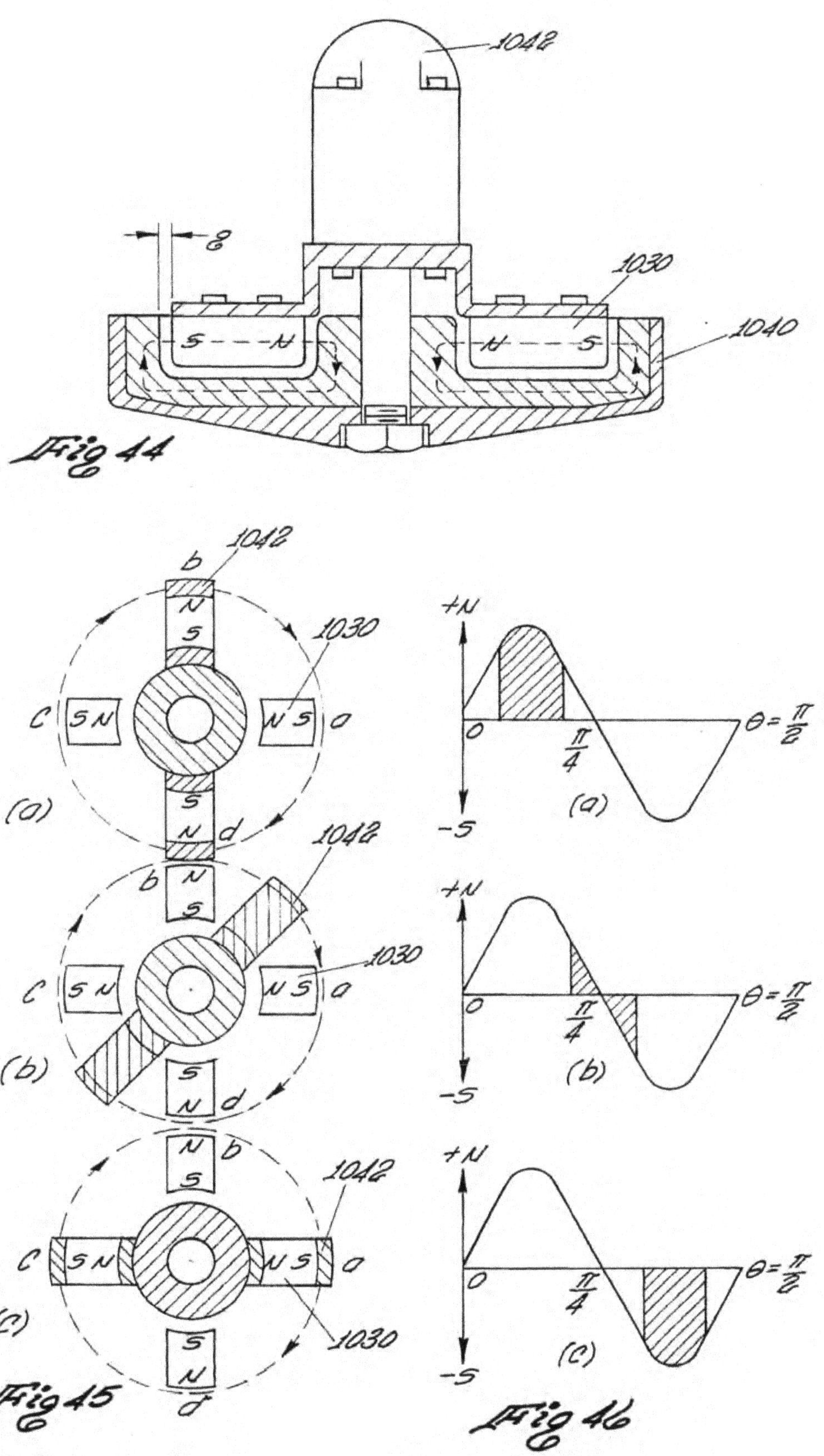

U.S. Patent May 12, 1987 Sheet 15 of 16 **4,663,932**

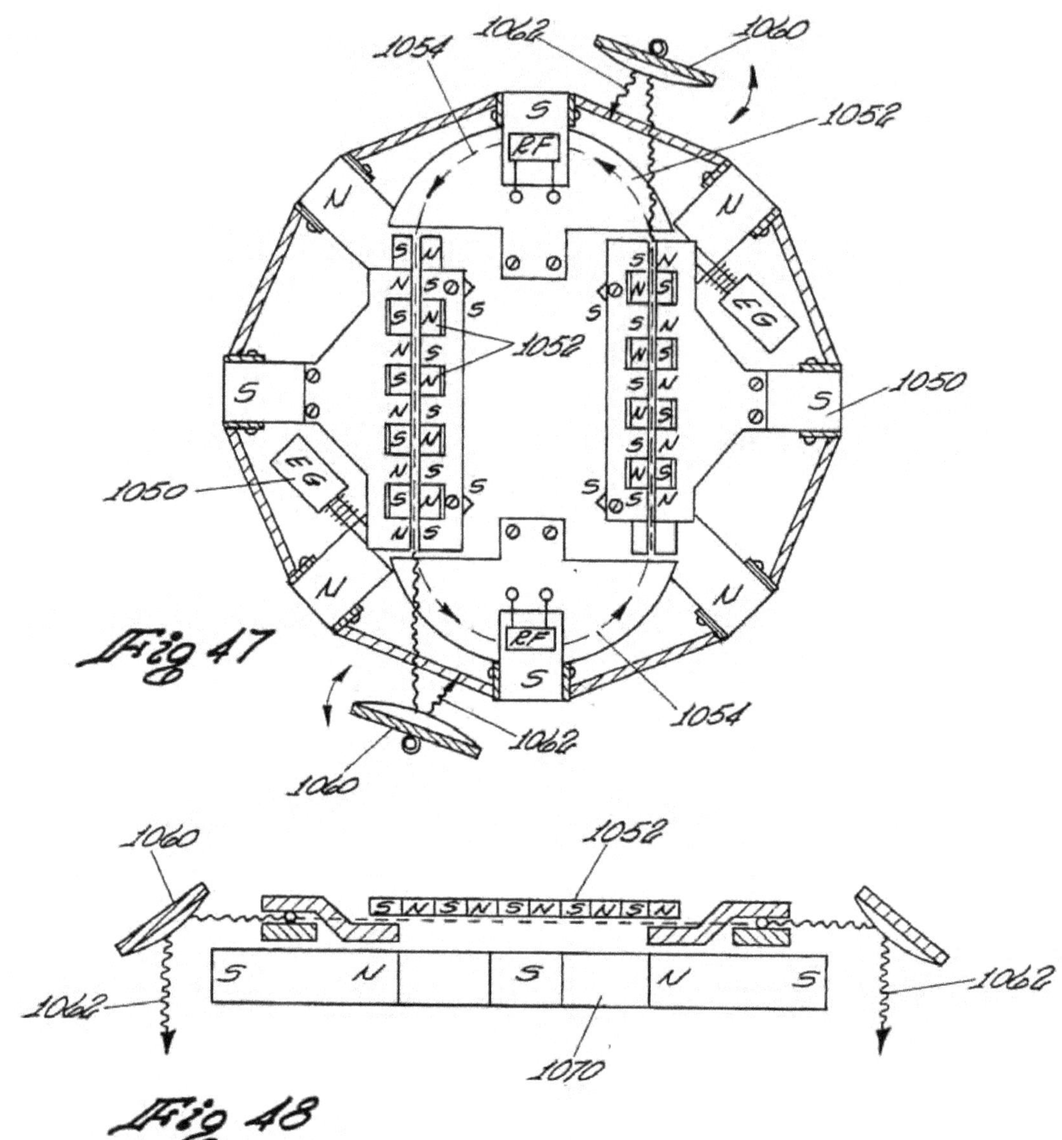

U.S. Patent May 12, 1987 Sheet 16 of 16 4,663,932

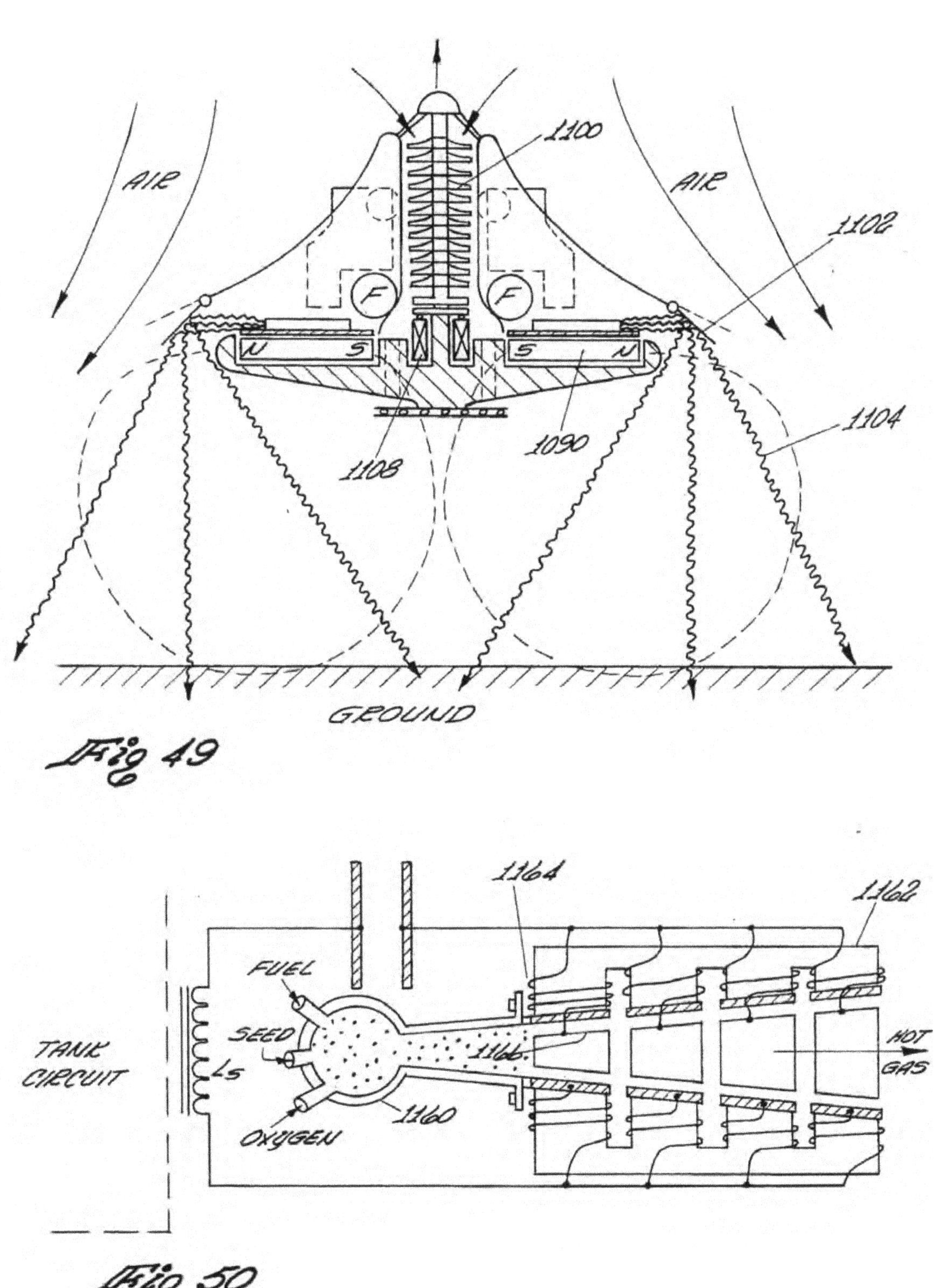

Fig 49

Fig 50

4,663,932

DIPOLAR FORCE FIELD PROPULSION SYSTEM

BACKGROUND OF THE INVENTION

1. Field of the Invention

The present invention relates in general to a system and method for producing a reactive force on an aerospace vehicle to cause rotation or vibration of dipoles of neutral particles having a selected electrical dipole characteristic and more particularly to a dipolar force field propulsion system for a aerospace vehicle utilizing a crossed electric E field and a magnetic B field for establishing a spatial force field region wherein a control means establishes a predetermined spatial and time relationship between the alternating electric field, alternating magnetic field and dipole rotation for a selected frequency to produce an reactive thrust.

2. Description of the Prior Art

In spacecraft propulsion systems, the use of chemical rocket engines which use combustion of chemical fuels to produce a large amount of thrust necessary to lift loads from the earth's surface is known. The term "thrust" is defined to mean the amount of propulsive force developed by a propulsion engine and is typically related to a rocket engine that is used for boosting a space vehicle from the earth's surface into orbit. The known space propulsion systems must have sufficient thrust to raise the spacecraft from the earth's surface and that thrust must be greater than the weight of the vehicle to be lifted from the earth's surface and placed into orbit.

Once the spacecraft has been boosted into space or orbit, the required spacecraft thrust is minimal compared to the thrust required for lifting the vehicles from the earth's surface.

When a spacecraft is in space or in orbit, it is desirable to have the ratio of thrust produced to the rate of consumption of the fuel to be high as possible and this is generally referred to as "specific impulse." In space or in orbit, a spacecraft propulsion system having a high "specific impulse" capability is highly desirable.

Thus, it is known in the art of space propulsion systems that the chemical rocket engines are capable of providing the requisite thrust necessary to lift large payloads from the earth's surface into orbit.

Once the spacecraft and its payload is in orbit, it is desirable for the spacecraft propulsion system to be able to change the orbit, speed and/or orbital position of the spacecraft with a "specific impulse" propulsive force.

A number of propulsion systems have the capacility of providing "specific impulse" thrust for changing the orbit, speed and/or orbital position of a spacecraft.

One such known propulsion engine is generally referred to as "electrostatic propulsion systems" wherein the thrust is created by electrostatic acceleration of ions created by an electron source in an electric field. Electrostatic propulsion systems have very high specific impulse but have limited thrust capacilities. Where an excessively large amount of thrust is required, the size and weight of the electrostatic propulsion systems become excessive. Examples of known electrostatic propulsion systems are disclosed in U.S. Pat. No. 3,866,414; U.S. Pat. No. 3,537,266 and U.S. Pat. No. 3,095,163. Electrostatic propulsion systems include electrostatic engines such as ion engines as evidenced by the above-described United States patents.

Another type of known space propulsion systems are generally referred to as "electric arc" engines. Electric arc engines or propulsion systems use an electric arc to heat a propulsion gas which is then passed to a standard rocket nozzle to provide thrust. Electric arc propulsion systems are capable of generating considerable amounts of thrust and have specific impulse thrust greater than those of chemical engines. However, the specific impulse thrust levels of electric arc engines are lower than the specific impulse thrust of electrostatic propulsion systems. Typical electrothermal or electric arc propulsion systems are disclosed in a book by Robert Jahn entitled "*Physics of Electric Propulsion*", McGraw Hill, 1968.

Another known type of spacecraft propulsion system is generally referred to as electromagnetic propulsion systems which includes magnetohydrodynamic (MHD) thruster or magnetoplasmadynamic (MPD) thruster. The MHD or MPD thrusters are capable of providing both high thrust density and high specific impulse. The MHD or MPD thrusters utilize a propellant gas which is ionized to form a plasma which is accelerated by magnetic and electric fields and is then passed through an expansion nozzle to provide thrust. In a MHD thruster or MPD thruster, the plasma is a body of gas which comprises a substantial number of free electrons and ions, but has an overall neutral electrical charge providing a plasma which is electrically conductive. The known MHD or MPD thrusters utilize the interaction of magnetic fields produced by electrical currents and conductors on the spacecraft with an electrically conductive environment to produce a reaction thrust. Several typical MHD thrusters or MPD thrusters are disclosed in U.S. Pat. No. 3,735,591; U.S. Pat. No. 3,662, 554; U.S. Pat. No. 3,535,586; U.S. Pat. No. 3,505,550; U.S. Pat. No. 3,371,490; U.S. Pat. No. 3,527,055; U.S. Pat. No. 3,343,022 and U.S. Pat. No. 3,322,374.

It is also known in the art to combine a jet propulsion power plant with a magnetoplasmadynamic generator to produce a hybrid propulsion system. One such propulsion system is disclosed in U.S. Pat. No. 3,678,306.

The use of a controlled fusion device which generates electrical energy utilizing an ionized gas plasma in a space propulsion system is disclosed in U.S. Pat. No. 3,324,316.

The design of plasma propulsion systems having special magnetic fields for controlling the specific impulse characteristics of the plasma propulsion device is disclosed in U.S. Pat. No. 3,191,092.

In addition to the above described space propulsion systems, the inventor of the present application published an article entitled "Electromagnetic Propulsion Without Ionization" which appeared in the AIAA/SAE/ASME 16th Joint Propulsion Conference which was held on June 13, 1980 to July 2, 1980 in Hartford, Conn. The paper presented at the above-described 16th Joint Propulsion Conference disclosed the concept of electromagnetic propulsion without ionization. Specifically, the paper disclosed that when an alternating electric field is applied to a polarized or polarizable material, the dipole of the material can be made to rotate at high frequency. If an alternating and synchronized magnetic field is supplied at right angles to the electric field, a Lorentz force is generated which propels the dielectric fluid without the necessity for ionization and the consequential energy losses arising from the ionization process. The thrust so generated is proportional to the polarization, the frequency of the dipole rotation and the magnetic field strength. The propellant selected for use as the polarizable material is characterized by having a high permanent molecular dipole movement-to-mass ratio and is accelerated by Lorentz forces to useful exit velocities. A spacecraft having the induced dipole electromagnetic propulsion system is accelerated by Newton's Third Law of Motion, or the reactive thrust principal.

SUMMARY OF THE INVENTION

The present invention relates to a novel, unique and improved dipolar force field propulsion system. In the prefered embodiment of the present invention, the dipolar force field propulsion system includes means for generating an alternating electric field having its electromotive lines of force extending in a selected direction. The alternating electric field varies at a selected frequency and has an electric field strength of a predetermined magnitude. A means for generating a rotating or alternating current magnetic field is provided with the electromagnetic lines of force extending in a direction which is at a selected angle relative to the selected direction of the electromotive lines of force. The electromagnetic lines of force cross and intercept the electromotive lines of force at a predetermined location to define a spatial force field region. The frequency of the alternating magnetic field is substantially equal to the selected frequency of the alternating electric field and has a predetermined phase angle therebetween. The magnetic field has relatively high flux densities in the order of a fraction of one tesla or more. The propellant material is a source of neutral particles of matter having stabilized, electrically induced or permanent dipoles having preselected internal breakdown characteristic which is greater than the magnitude of the applied electric field. The dipoles of the matter are capable of being driven into controlled rotation at the selected frequency by the alternating electric field and crossing the alternating electromagnetic field. A means for vaporizing the matter into a gaseous state yet below the thermal ionizational level thereof and for transporting the vaporized material in the gaseous state into the spatial force field region which is defined by the crossed electromotive lines of force and electromagnetic lines of force. The alternating cross field formed by the electromotive lines of force and the electromagnetic lines of force cause the dipoles to rotate at the selected frequency and to produce an acceleration force which is substantially normal to the plane of the electromotive and the electromagnetic lines of force to produce a reactive thrust. A control means which is operatively coupled to the means for generating the alternating electric field and to the means for generating an alternating magnetic field and which is responsive to the dielectric properties of the vaporized matter located in the spatial force field region having a well-defined relation between the electric field, electromagnetic field and dipole orientation for any selected frequency.

The known prior art space propulsion systems have inherent limitations in terms of providing sufficient thrust based upon the mass and weight of a propulsion system on the earth's surface in order to lift a spacecraft from the earth's surface and to place the same into orbit or space. The primary limitation can be characterized specifically by the mass of propellant required, by weight, to the mass of payload to be placed into space. Known spacecraft propulsion systems utilizing a chemical engine generally require propellants wherein the aggregate weight of the propellant is twenty to thirty times the aggregate weight of payload to be lifted from the earth's surface and to be placed into orbit.

The known electrostatic propulsion systems or ion propulsion systems and the electric arc propulsion systems are limited to operation in the vacuum of space and provide satisfactory high "specific impulse" thrust but are unsatisfactory for providing a substantial amount of thrust as required for liftoff of a spacecraft. In order to generate sufficient thrust for lifting of a payload from the earth's surface into orbit, the size, weight and complexity of the spacecraft propulsion systems limit the desirability of using the same in such a spacecraft and to provide the necessary "specific impulse" thrust required for changing orbital speed, direction and/or position.

In the known MHD or MPD propulsion systems, it is necessary to provide sufficient energy in order to ionize the propellant. The energy required to ionize the propellant, which is typically easily ionizable gas, reduces the overall efficiency of the propulsion systems and requires substantial cooling systems in order to obtain the proper operating conditions to increase the reliability and lifetime of such propulsion systems.

In the known MHD propulsion systems, it is necessary to include a seeding propellant which is injected into the hot gases wherein the seeding material is generally a low ionization potential compound such as, for example, potassium or cesium.

The present invention overcomes the inherent limitations and problems associated with the known spacecraft propulsion systems.

One advantage of the present invention is that a unique, novel and improved dipolar force field propulsion system utilizes a propellant in the form of a vaporized gaseous matter which is in an unionized state. The reactive thrust can be developed by controlling the operating characteristics of the crossed alternating electric field and alternating current magnetic field which defines the spatial force field region adapted to have the vaporized polarizable material, which is not ionized, transported thereto.

Another advantage of the present invention is that the electronic excitation level of the polarizable dipole material can be increased either prior to or after the vaporization thereof into a gaseous state to improve the operating efficiency of the dipole force field propulsion system.

A yet further advantage of the present invention is that a means are provided for generating a reactive thrust which is adapted for propelling a spacecraft from the earth's surface, into orbit and subsequently into space wherein the initial thrust and specific impulse can be provided which are equal to or greater than those provided by the known spacecraft propulsion systems.

A still yet further advantage of the present invention is that a unique and novel method for propelling a spacecraft with a reactive thrust derived from using a propellant comprising neutral particles of matter having an electric dipole characteristic and a breakdown characteristic which is greater than the magnitude of an applied electric field.

A still yet further advantage of the present invention is that the phase angle between the alternating electric field and the alternating magnetic field can be varied so as to control the magnitude of the reactive thrust produced by the rotation of the dipoles of material.

A still yet further advantage of the present invention is that a unique and novel spacecraft having a "X-wing" configuration which includes means for exciting the energy level of the polarizable or dipole material to an excited level wherein the excited atoms of material when used as a propellant is capable of rendering both thrust and specific impulses of thrust at controlled levels which is directly proportional to the excited state of the gaseous material.

A still yet further advantage of the present invention is that the propulsion efficiency of the inductive dipolar force field propulsion system increases as a function of mass ratio and can approach acceptable operating efficiencies.

A still yet further advantage of the present invention is that the excitation power can be a microwave source having a selected frequency which can be located either internal or external to the spacecraft. Under certain idealized conditions, the frequency of the microwave radiation source can be precisely selected relative to the frequency of rotation or absorption characteristics of the dipole material such that substantially all of the microwave radiation transmitted to the spacecraft from an external source can be fully absorbed without reflecting any part thereof.

A still yet further advantage of the present invention is that a MHD electric power generator can be utilized on board of the spacecraft to generate the electrical energy required to produce the electric and magnetic field which is utilized to establish the spatial force field area for producing the reactive thrust from the interaction of the crossed electric field and magnetic field on the induced dipole material occupying this region.

A still yet further advantage of the present invention is that cryogenic cooling of superconductive magnets can produce extremely high, dense magnetic fields in the order of one tesla or more. By controlling this field strength as well as the switching rate or frequency of the magnetic fields, both the efficiency of the dipole propulsion system and the amount of thrust produced can thereby be determined.

A still yet further advantage of the present invention is that a electromagnetic propulsion system utilizing the teachings of this invention can produce in the order of 10^6 pounds of thrust level using known or anticipated power sources and known superconductive magnetic materials.

A still yet further advantage of the present invention is that a shuttle aircraft can be designed utilizing a hybrid propulsion system wherein the lift and thrust are accomplished by aerodynamic, electromagnetic and chemical rocket propulsion systems so as to exploit the characteristics of each system at an optimum time during trajectory of spacecraft travel.

A still yet further advantage of the present invention is that the spacecraft propulsion system disclosed herein is capable of utilizing the earth's atmosphere as a propellant having an appropriate excitation level required in order to initiate the polarization dipole reactive thrust generation for purposes of lifting a spacecraft from the earth's surface into orbit. Once the spacecraft has been propelled into orbit and then into deep space, the dipole force field propulsion system is capable of utilizing matter in interstellar space as a propellant without the necessity of ionizing the same in order to develop the reactive thrust necessary to propel a spacecraft into deep space.

A still yet further advantage of the present invention is that the dipolar force field propulsion system provides a method for accelerating neutral particles of matter without the creation of an ionized or plasma state. As a result, a force density can be established in a gas over a large distance without the restriction of skin depth or Debye lengths. This property, in addition to the recycling of excitation radiation and rebounding collision processes, offers the potential for the creation of a class of more efficient propulsion systems for aerospace vehicles.

A still yet further advantage of the present invention is that the dipolar force field propulsion system operates at lower jet velocities at large volumetric mass flow rates. Therefore, greatly reduced noise levels are possible. The field extends beyond the structure of the aerospace vehicle itself to move the mass and thereby permits operation in more rarified environments, such as higher altitudes, where pressures and temperatures are lower, permitting high Rydberg excitation states to exist.

A still yet further advantage of the present invention is that the aerospace vehicle's structure can be designed such that electronic control of thrust direction can be achieved which can be changed instantly with the flick of a switch. The use of electronic switching can provide increased maneuverability and faster response reaction times. Further, electric power can be provided to the aerospace vehicle by super conductive radio frequency generators or by the process of magnetohydrodynamics, or by beamed power from ground or orbiting power stations. The existance of an excited gas field around the vehicle can be used in absorbing offending external microwave beams as well.

A still yet further advantage of the present invention is that it appears that the ejection of electromagnetic momentum will provide for some capability of producing a small thrust in the vacuum of space itself.

A still yet further advantage is that the apparatus and method disclosed herein can be used for accelerating particles of matter and have wide potential applications for isotope separations, particle beam devices, chemical accelerators, nuclear devices, molecular beam devices and the like.

BRIEF DESCRIPTION OF THE DRAWINGS

These and other objects and advantages of the invention, together with the various features and advantages, can be readily understood from the following more detailed description of the prefered embodiment taken in conjunction with the accompanying drawing in which:

FIG. **1** is a diagramatic representation of electrodes for establishing an alternating electric field in the presence of a alternating magnetic field to define a spatial force field region for inducing rotation of a dipole to produce a reactive thrust;

FIG. **2** is a vector diagram of the Lorentz forces acting on each charge at the end of a dipole;

FIG. **3*a*** is a diagramatic representation of the elliptical orbit generated by an electron relative to its nucleus showing the aphelion point and the perihelion point of the orbits;

FIG. **3*b*** is a graph of the charge density of the atom plotted as a function of electron distance in Bohr radii (R_o) depicting the variance in charge density as a function of radius of the orbit;

FIG. **4** is a plot of the electronic energy levels of hydrogen gas as a function of the principal quantum number (n) of an excited hydrogen atom;

FIG. **5** is a graph of the polarizability of an atom at various levels of excitation and reduced ionization potential and depicting the excitation frequency and breakdown voltage of the dipole material;

FIG. **6** is a graph showing the particle accelerations which can be obtained for dipolar molecules in a plurality of excited states;

FIG. **7** is a diagramatic representation of a simplified dipolar force field propulsion system utilizing the teachings of the present invention;

FIG. **8** is a schematic representation of one embodiment of the present invention having a plurality of stages, each of which have linear spatial force field regions and utilizing a lasar as a source of excitation of the gas, and cryogenic cooling to increase the efficiency of the dipolar force field propulsion system;

FIG. **9** is a diagramatic representation is one view of one of the stages of the dipolar force field propulsion system illustrated in FIG. **8**;

FIG. **10** is an electrical schematic diagram of the electrical component connections which includes therewith a representation of the capacitance effect of the vaporized propellant located in the spatial force field region;

FIG. **11** is a graph of a specific impulse versus operating perimeters for various dipole moment/mass ratios;

FIG. **12** is a graph of the thrust/power ratio versus velocity for the propellant in the vaporized and unionized state;

FIG. **13** is a pictoral representation of dipolar force field propulsion system having an elongated rectangular channel having a spatial force field region between plates for establishing the electromotive lines of force and wherein the magnetic lines of force of the magnetic field cross the electromotive lines of force of the electric field within the spatial force field region and wherein the vaporized gas is first passed through an excitation source which raises the electronic energy level thereof to a substantially higher level and wherein the excited atoms deactivate or decay to a ground state producing emission as the propellant passes through an outlet nozzle and the emissive radiation so generated is fed back through a mirror reflective system back to the input excitation source;

FIG. **14** is a pictoral representation, partially in sectional view, showing a high frequency torroidal dipolar force field propulsion system utilizing the teachings of this invention;

FIG. **15** is a front end view partially in cross section showing the construction of the various structural members which define a torroidal shaped spatial force field region;

FIG. **16** is a graph representing the mechanical efficiency plotted as a function of mass ratio of the atoms utilized as the dipolar propellant matter relative to vehicle mass;

FIG. **17** is a graph of range of force field plotted as a function of decreasing medium gas density for a number of different mass ratios;

FIG. **18** is a diagramatic representation, in cross section, showing the details of wing construction of an aerospace vehicle showing in particular the structure of the magnetic field and electric field for establishing an external spatial force field region using atmospheric gas as the propellant;

FIG. **19** is a top plan view illustrating a method of thrust directional control employing segmented electrically conductive plates and a switching mechanism for the wing construction of FIG. **18**;

FIG. **20** is a schematic diagram of the equivalent circuit of the wing illustrated in FIG. **19**;

FIG. **21** is a top view, partially in cross section, of a discoid shaped vehicle having a rotating nuclear bed reactor and a single wing showing the construction thereof adapted to provide an external force field region;

FIG. **22** is a side view, partially in cross section of the discoid shaped vehicle of FIG. **21**;

FIG. **23** is a front plan view, partially in cross section, of the discoid vehicle illustrated in FIG. **21**;

FIG. **24*a*** is a simplified electrical schematic diagram showing the internal and external capacitive arrangement of the discoid vehicle of FIG. **21**;

FIG. **24*b*** is a simplified electrical schematic diagram showing the internal and external capacitive inductive elements of the discoid vehicle of FIG. **21**;

FIG. **25** is a graph illustrating the microscopic collisional processes between excited and ground state dipolar atoms forming the propellant matter;

FIG. **26** is a graph of the comparative propulsion efficiency of three known spacecraft systems versus the relative vehicle velocity of the spacecraft;

FIG. **27** is a graph of the body force developed in a gaseous atmosphere plotted as a function of the magnetic field frequency for several different altitudes;

FIG. **28** is a graph of the body force plotted as a function of the magnetic field times frequency product for various levels of excitation states of a vaporized gas utilized as a propellant;

FIG. **29** is a front plan view of an "X-wing" spaceshuttle aircraft utilizing the teachings of the present invention;

FIG. **30** is a top plan view, partially in section, showing the "X-wing" shuttle spacecraft of FIG. **29**;

FIG. **31** is a pictoral representation, partially in section, showing the details of the construction of the upper and lower wing of the "X-wing" shuttle spacecraft of FIG. **29**;

FIG. **32** is a sectional view taken along section lines **32**—**32** of FIG. **31**;

FIG. **33** is a schematic diagram showing the inductance and capacitance of the wings of the "X-wing" shuttle spacecraft of FIG. **29**;

FIG. **34** is a schematic diagram of an alternating current power source for supplying electrical power to the inductive and capacitive components of the aircraft of FIG. **29**;

FIG. **35** is a side view, partially in cross section, showing a two stage inductive dipolar force field propulsion system;

FIG. **36** is a front plan view of the two stage inductive dipolar force field propulsion system of FIG. **35** showing the spiral coil winding in detail;

FIG. **37** is a simplified block diagram showing the overall electrical power circuit for the inductive dipolar force field propulsion system of FIG. **35**;

FIG. **38** is a diagramatic representation partially in cross section of a vertical takeoff and landing vehicle (VTOL) using the inductive dipoler force field propulsion system;

FIG. **39** is a partial top plan view of the VTOL spacecraft illustrated in FIG. **38**;

FIG. **40** is a pictoral representation partially in section showing a means for controlling the region of the excitation of gas molecules in the atmosphere beneath the VTOL spacecraft to bring about thrust and direction control;

FIG. **41** is a top plan view of the VTOL spacecraft illustrated in FIG. **40**;

FIGS. **42***a*, **42***b* and **42***c* depict the effect of controlling the excitation source for increasing the level of excitation of gas atoms in the atmosphere in the vicinity of a VTOL vehicle to provide thrust for causing the vehicle to be lifted and directionally controlled from earth, and adapted to be turned to the right or to be turned to the left, respectively;

FIG. **43** is an isometric view showing a means for producing an alternating magnetic field using D.C. superconductive magnetic coils;

FIG. **44** is a pictoral representation partially in cross sectional view, showing fixed magnets in a rotating ferrite slotted disc;

FIGS. **45***a*, **45***b* and **45***c* are a series of pictoral representations showing the ferrite rotor in various angular positions relative to the magnets;

FIGS. **46***a*, **46***b* and **46***c* are graphs showing the resulting field polarity and magnitude with the ferrite rotor in various angular positions as illustrated in FIGS. **45***a*, **45***b* and **45***c*, respectively;

FIG. **47** is a top plan view of the magnetic configuration of a spacecraft utilizing the inductive dipolar force field propulsion system of the present invention as a means for generating a reactive thrust adapted for propelling a spacecraft utilizing a wiggler magnet arrangement as a means for accelerating an electron beam and producing a controllable continuum of vacuum ultraviolet radiation for excitation of the ambient gaseous atoms to an electronic excited state;

FIG. **48** is a pictoral representation of the front plan view of the magnetic configuration illustrated in FIG. **47**;

FIG. **49** is a pictoral representation, partially in cross section, showing an embodiment of a VTOL vehicle which is adapted to utilize gaseous atoms in the atmosphere as a propellant and for exciting the same with a source of radiant energy in order to cause the VTOL to hover near the earth's surface; and

FIG. **50** is a diagramatic representation of an MHD plasma energy source having pumped mutually coupled LCR circuits which is adapted for use in the "X-wing" spaceshuttle illustrated in FIG. **29**.

DESCRIPTION OF THE PREFERRED EMBODIMENT

Before commencing with a detailed description of the preferred embodiment and alternate embodiments, a brief description of the electrodynamics of moving media particularly with respect to a model of a dipolar fluid will be first considered.

A description of the model of a dipolar fluid and the resulting equations developed by a force acting on the dipolar fluid is set forth in a book entitled *Electrodynamics of Moving Media* by Paul Penfield, Jr., and Hermann A. Haus which is published as Research Monograph Number 40 by the MIT Press, Cambridge, Mass. at pages 47 through 53. As stated in the description of the model of a dipolar fluid in the above-described Penfield and Haus reference, in a uniform field, the force density can be defined by the following formula:

$$f_k = \dot{P} \times B \quad (1)$$

wherein

f_k=Force Density (Newtons/Cubic Meter)

$\dot{P}$=Polarization Current Density of Dipolar matter (in A/M^2); and

B=Magnetic Field Induction (Tesla)

From the formula identified as equation (1) above, the force density is a function of the polarization current of the dipolar material times the magnetic field intensity. Polarization is defined, for purposes hereof, as the average electric dipole moment per unit volume. The derivative thereof with respect to time yields current density.

Experiments have been conducted to verify that mechanical forces can be developed based upon the above-described formula and the results of such experiments were disclosed in an article entitled "Mechanical Forces of Electromagnetic Origin" by G. B. Walker and G. Walker of the Electrical Engineering Department, University of Alberta, Edmonton, Canada, which was published in a periodical entitled "Nature" at Volume 23, Sept. 30, 1976. The experiments disclosed that the above-identified formula results in a reactive force being generated.

Equation (1) above is a compact mathematical expression which represents the underlying microscopic physical forces taking place at the atomic level. This understanding is essential in order to appreciate and understand the teachings of the present invention.

Referring to FIG. **1**, a pair of electric dipoles **100** are shown consisting of oppositely charged ends, **102** and **104**, end **102** being the positively charged end and end **104** being the negatively charged end. The two dipole ends **104** and **106** are displaced a fixed distance "s" apart from each other and are free to rotate about an axis **106** which is the positively charged end **102**. The dipoles are shown pictorially to be an elongated shaft terminating in a sphere at each end thereof with the charges concentrated at each end thereof. In fact, in an actual ground state atom, the electrons exist as a cloud shifted from the nucleus.

As illustrated in FIG. **1**, the dipoles are situated in a crossed electric field and magnetic field referred to in the art as a Lorentz field. The electric field can be generated by a means for generating an alternating electric field having its electromotive lines of force extending in a first or selected direction. The alternating electric field varies at a selected frequency and the electric field is selected to have an electric field strength of a predetermined magnitude. In the preferred embodiment, the magnitude of the electric field is less than the known characteristic field ionization of particle or particles of matter having the dipole formed therein.

FIG. **1** includes a means for generating an alternating magnetic field having its magnetic lines of force extending in a second direction which is at a predetermined angle, which in the preferred embodiment, is at 90°, to the first or selected direction of the electromotive lines of force defining the electric field. The pole face of the magnet is shown as **122**. The magnetic lines of force intercept the electromotive lines of force at a predetermined location to define a spatial force field region. The frequency of oscillation of the alternating magnetic field is substantially equal to the selected frequency of the alternating electric field. Also, the oscillation of the alternating magnetic field magnitude is at a selected phase angle with the alternating electric field. As will be developed further herein, the magnetic field has a flux density which when multiplied times the selected frequency produces a Tesla-Hertz level which is less than the selected field ionization limit of a particle placed into the field.

In FIG. 1, the electric field shown by dashed lines **114**, is generated by a pair of electrodes **116** and **118** with electrode **116** being the positively charged electrode and with electrode **118** being negatively charged at an instant of time. The voltage applied to the electrode cyclically varies as a cosine function, Cos (wt). The magnitude of the electric field is chosen so as not to cause electrical breakdown of the dipole, that is to cause separation of the opposite ends of the dipole from each other. If the magnitude of the electric field is less than the electrical breakdown of the dipole, the electrons remain bound to each other at a fixed distance "s" apart. Likewise, a magnetic field B, shown by vectors **112**, is applied to the dipoles. Preferably, the magnetic field has a flux density which is as intense as is practically possible based upon the frequency of the alternating magnetic field and the Tesla-Hertz level thereof relative to the selected field ionization limits of the particle of matter subjected to the force field. The magnetic field applied to the dipoles varies as a sine function, Sin (wt).

Both electrode pair **116** and **118** and the magnetic field **112** are controlled to establish a predetermined spatial and time relationship at the selected frequency of the alternating electric field, the alternating magnetic field and the ultimate dipole rotation orientation.

When the electric field E is initially applied to dipoles **100**, the dipoles **100** will experience a torque that will twist them into an orientation such that they are parallel to the electric field lines **114** with the opposing charges facing each other at a given electrode. The dipole may rotate in either a clockwise or counterclockwise direction depending on its initial position. However, as will become apparent, the direction of rotation is immaterial to the translatory forces that are to be generated on the dipoles **100** as a whole. If an alternating electric field, E, is applied to the electrodes **114** and **116**, the dipoles **100** can be made to rotate or oscillate about its center mass, which is generally the positively charged end of the dipole. The frequency of rotation is in the megacycle range and the dipoles' rotation follows the frequency of the electric field. Thus, the dipole is driven into cyclic motion, which may be rotational or vibrational, by the electric field. When the alternating magnetic field is imposed on the dipoles, forces are exerted on each charge of the dipole given by the following Lorentz equation:

$$F = qv \times B \qquad (2)$$

where

q=charge on each end of dipole (coulombs);

v=tangential velocity of each charge (m/s); and

B=magnetic field in teslas.

As shown in FIG. 2, the force acts in a direction perpendicular to the plane of the electric and magnetic fields, which is along the X axis in FIG. 2. For velocity components colinear with the magnetic field lines, which is along the Y axis in FIG. 2, no force is produced in the X plane since the cross-product of the velocity and magnetic field is equal to 0. Only velocity components perpendicular to the magnetic field generates forces in the X and Z plane. The forces that are generated as the dipole is rotated through each quadrant in FIG. 2 can be summarized by analyzing equation (2) at each quadrant location and a chart thereof as set forth hereinbelow.

TABLE 1

		Forces on Negative Charge (Clockwise Rotation)				
Quadrant Location	Q	B(Y)	V(x)	V(z)	F(z)	F(x)
I	−e	0	+wR	0	0	0
II	−e	−MAX(Y)	0	+wR	0	+qVB
III	−e	0	−wR	0	0	0
IV	−e	+MAX(Y)	0	−wR	0	+qVB

As is apparent from the above chart, in respect to the negative end, at quadrant location I, the B field is 0 and the voltage in the z direction is 0 and the velocity in the x direction is equal to (+WR). Thus, applying the equation (2) to the above values, the force in the x direction and the z direction are both 0.

At quadrant location II, the B field is at a maximum negative designated as −MAX(Y), the velocity in the z direction is equal to +wR and the velocity in the x direction Vx is equal to 0. Applying the force equation, a force equal to a +qVB is produced causing the dipole to be forced to the right.

At quadrant location III, the same conditions exist as in quadrant I and the force is equal to 0, as both fields reverse direction.

At quadrant IV, the B is equal to a +MAX(Y), Vz is equal to −wR and Vx is equal to 0. Thus, the force in the x direction is also equal to a +qVB.

For the positive end of the dipole, the sign of charge is now positive, but its velocity is also reversed, since by Newton's third law, it moves opposite the direction of the negative end. Thus, the net force along the X-axis is the same.

The same analysis would apply to the second dipole, being noted that the second dipole is shown rotating in an opposite direction but the Z velocity components are the same for each charge. The dipole rotation can be commenced in either direction based upon the probability of the location of the electron at the time of the application of the electric field thereto.

The electric forces (E) for the negative charge on each dipole vary as a cosine function yielding a velocity which is its integral or sine function. Thus, the net force is vector sum of the forces on the negative and positive charges:

$$F = +q\,(V \sin wt)\,(B \sin Wt) + (-q)\,(V \sin wt + \pi)\,(B \sin wt) = 2q\,w\,RB \sin^2 wt. \qquad (3)$$

Since 2qR is the dipole movement (p=qs), the net force on each dipole is shown in equation (3):

$$F = p\,B\,w\,\mathrm{Sin}\,^2wt. \qquad (4)$$

The average force is found by integrating equation (4) over a complete cycle and dividing by (2π):

$$F = \tfrac{1}{2}\,pBW. \qquad (5)$$

For purposes of this invention, the term "particle" is intended to cover an atom of matter, a molecule of matter or a colloid of matter which can be defined as an aggregate of molecules stuck together. As an example, consider the case where the particle is water. A water molecule H_2O has the permanent dipole movement equal to 1.85 Debyes (a Debye is equal to 3.3×10^{-30} Coul-meter) due to the assymetry of the hydrogen bonds with the respect to the oxygen atom. In addition, an induced dipole movement P_i can be created when an electric field is applied given by the following equation:

$$P_i=\epsilon_o\alpha\ E \qquad (6)$$

where

ϵ_o is the permittivity constant; and

α is the polarizability (m^3).

Polarizability has the dimensions of volume, and a value that approximately corresponds to the actual volume of the atom or molecule. The volume of a molecule can be increased significantly (and hence its polarizability) by exciting the particles' outer electrons to high energy levels. The radius of a quantum orbit in a simple Bohr atom increases with the square of the principal quantum number (n). Hence, the polarizability increases as the volume by the following equation (40):

$$\alpha=4/3\ \pi\ n^6\ R_o^3 \qquad (7)$$

In order to aid the explanation of the polarization of an atom, the subject shall be treated in a classical manner and should be based upon a reference to a simple Bohr atom (hydrogen) with a single proton at the core. The electron is assumed to have been excited to a higher energy state, and is in orbit about the nucleus as shown in FIG. **3**. An energy level diagram thereof is shown in FIG. **4** and will now be described in detail.

FIG. **3***a* is a graph showing the orbit traversed by an electron **128** of a hydrogen atom having a proton **130**. The atom is in a highly excited state. The electron (**128**) traverses a path shown by arrows **132** and the distance between the electron **128** and the proton **130** is shown by "r." The shortest distance between the electron **128** and the proton **130** is shown by "r_p," the lowest orbit point being the perihelion. The greatest distance between the electron **128** and the proton **130** is shown as "r_a" (the highest orbit point being the aphelion).

For large (n), the Rydberg electron moves in a nearly hydrogenic orbital around a core which consists of an atomic ion. This illustration shows a classical Bohr orbit. In reality, the electron is viewed as a cloud of charge. Hence, the charge in any region is equal to the volume of that region times the charge density. The average charge density is proportional to the time the electron spends in that region of its orbit. The faster the electron moves through a region, the less time it spends in that region and, therefore, the less average charge in that region. Classically, the charge density varies inversely as the speed of the electron. In FIG. **3***a*, as the electron moves further from the nucleus, the slower its speed, and hence a larger concentration of charge at a distance from the core. Hence, the Rydberg atom has an electric dipole moment, particularly when an external electric field is applied to the particle. In the simplest view, this moment is equal to the product of electron's charge times the distance from the ion core:

$$p=e\ n^2R_o \qquad (8)$$

where R_o is the ground state radius of the electron. For n=20, in the case mentioned earlier, $p=1.6\times10^{-19}$ (400) $(10^{-10})=6.4\times10^{-27}$, coul-meters, more than 1939 Debyes, 1048 times larger than H_2O! The dipole moment-to-mass ratio for a simple excited hydrogen atom is thus nearly equal to unity (one). Hence for a magnetic field of ½ Tesla, the acceleration corresponds to the value of the frequency, i.e., $10^6 m/s^2$ at one megacycle, etc. However, the induced electric field may be sufficient to ionize the atom as the atom or molecule is excited to higher and higher energy levels, it becomes more easily ionized. The ionization potential decreases inversely with the square of the principle quantum number:

$$U=U_i/n^2$$

The application of an external electric field E and magnetic field B distorts the path traversed by the electron **128** and pulls the electron to one side of the proton **130**. The effect of the external electric field E is to apply a moment onto the dipole in accordance with Equation (6).

FIG. **3***b* is a graph showing the charge density of the atom of hydrogen illustrated in FIG. **3***a* as a function of the distance of the electron **128** from the proton **130** in Bohr radii (R_o). As shown in FIG. **3***b*, when the electron is at distance "r_p" the charge density is high due to the close proximity of the electron **128** to proton **130**, even though the dwell time is short the charge density decreases as the distance "r" increases until the distance "r_a" is reached. At that point, the electron essentially reverses direction and the variance in speed results in a momentary increase in charge density.

As noted in Equation (8), the dipole moment p increases as the square of the dipoles energy level "n," wherein "n" is the quantum number of the energy level.

FIG. **4** is a graph of the effect of exciting hydrogen gas to various quantum levels "n" plotted as a function of electron volts (eV). The energy level of the hydrogen gas can be increased by means of a laser source or other energy source which is capable of raising the excitation level to a high quantum level. The Bohr radii increases as a square of the quantum number "n." For example, if n=2, the radius is four (4) times larger. The volume of the atom increases as a function of r^3, or N to the sixth (6th) power.

Thus from a theoretical aspect, one significant and important part of this invention is the increased operating efficiency and increased thrust that is obtained by exciting the atoms of the gaseous material to a high level of electronic excitation (sometimes referred to as a Rydberg atom). The relationship between the acceleration of dipolar particles in both a ground state and in an excited state and the effect thereof on the dipolar force field propulsion system can now be assessed. The ideal operational conditions of an inductive dipolar force field propulsion system can be developed as follows:

The particle acceleration has been derived earlier [equation (5)]:

$$\ddot{X}=\frac{1}{2}\ \frac{P_e}{M_o}\ (BW) \qquad (9)$$

The dipole moment (P_e) is that induced due to an applied electric field (E), to an excited atom:

$$P_e=\epsilon_oK_1n^6R_o^3E \qquad (10)$$

where ($n^6\ R_o^3$) is the polarizability in cubic meters, incorporating the recent evidence that the polarizability scales as n^7 for excited atoms. Here R_o is the Rydberg electron orbit radius for the ground state (n=1 for light elements), and K_1 is a correction factor of the actual ground state polarizability versus the actual atomic volume. If the electric field is too high, field ionization of the atom will occur; this limiting field (E_f) is given by the Coulomb equation:

$$E_f = \frac{ZKe}{R^2} \quad (11)$$

where (R) is the electron orbit radius, equal to:

$$R = n^2 R_o \quad (12)$$

and (Z) is the atomic number, and K has the value 9×10^9.

For any simple atom, the number of protons equals the number of neutrons in the nucleus, and thus the atomic mass is approximately:

$$m_o = 2Z\,M_p \quad (13)$$

where (M_p) is the proton rest mass. The maximum dipole moment-to-mass ratio is thus (combining equations (10), (11), (12) and (13):

$$\Omega = \frac{Pe}{M_o} = \frac{\epsilon_o K K_1 n^3 R_o e}{2M_p} \quad (14)$$

Note that (r) is apparently independent of (Z). We can evaluate this result by letting:

$\epsilon_o = 8.85\times10^{-12}$

$K = 9\times10^9$

$K_1 = 1$

$R_o = 0.5\times10^{-10}$M

$M_p = 1.67\times10^{-27}$ kg

$e = 1.6\times10^{-19}$ Coul

The result is:

$$r = 2\times10^{-4} n^3 \quad (15)$$

Consider the following examples:

For:

n=17, r=1

n=36, r=10

n=79, r=100

In order to obtain high Rydberg states (n>10), the gas should be cooled to reduce the chances of collisional quenching:

$$\underset{\text{(molecular energy)}}{3/2\,KT} < \underset{\text{(ionization energy)}}{Ui/n^2}$$

where (U_i) is the ground state ionization potential, and here (k) is Boltzman's constant and (T) is the temperature in degrees Kelvin. High n's are possible in thruster applications where selected propellants are utilized. A cryogenic gas such as, for example, the boil-off of liquid helium at about 5° K. may be used, thus a possible maximum (n) value is:

$$n_{max} = \left(\frac{U_i}{\frac{3}{2}kT}\right)^{\frac{1}{2}} = 141$$

In an inductive dipolar accelerator, described later in reference to FIG. 36, the acceleration is given by:

$$\ddot{x} = \frac{\epsilon_0 \alpha}{2M_o} R_c (BW)^2 \quad (16)$$

where R_c=the coil radius or field gap used in the magnet. We can calculate the limiting B-field frequency product before ionization is induced:

$$BwR_c < \frac{Zkl}{R^2} \quad (17)$$

$$\therefore Bw \approx \frac{Zkl^2}{R_c R^2} \quad \text{(Limiting Field-Frequency)}$$

Combining equations (16) and (17):

$$\ddot{X}_{max} = \frac{\epsilon_o K_1 K^2 Z e^2}{4 M_p R_c n R_o}$$

Evaluating this with k_1=1, and assuming R_c=1 cm, we obtain:

$$\ddot{X}_{max} = \frac{10^{10}}{n}\ \text{M/sec}^2 \quad (18)$$

For n=100, the acceleration is $X=10^9 m/s^2$, comparable to conventional electric and plasma thrusters. This is achieved at a field-frequency product of:

$$B\nu = \ddot{X}/2\pi r = 0.8\ \text{MHz-T} \quad (19)$$

Thus, assuming we can have high Rydbergs, at a magnetic field-frequency product of less than 1 MHz-T, the particle acceleration is comparable to conventional thrusters. The lifetime (τ_e) of the excited Rydberg atom is greatly increased at large values of n, in fact it scales as:

$$\tau_e \sim n^3 \quad (20)$$

(neglecting collisions and field effects). Hence, the lifetime can be long enough to be accelerated over the channel distance before deactivation:

$$\tau_e > L/V_g \quad (21)$$

where (L) is channel length and (V_g) is gas velocity. The Lorentz forces exerted on the excited Rydberg electron by the external B-field becomes comparable to the Coulomb forces holding the electron captive to the nucleus:

$$qVXB \approx \frac{ZKe^2}{R^2} \quad (22)$$

This can be made into a squeezing force to be used to minimize the chances of ionization at the cyclotron frequency ($w_c = eB/M_e$). Operation at lower pressures would also be desireable to reduce again the effects of collision frequency and increase the mean free path comparable to the size of the accelerator channel. In any event, any collisions that do take place should satisfy the following condition:

$$3/2KT \neq n(\nu_n - \nu_{n-1}) \quad (23)$$

That is, the collision energy should not correspond to any transition of either particle (vibrational, rotational or electronic). Finally, the conductivity (6) of the gas (degree of ionization) must be not so high that the skin depth (8) gets too low and the field does not penetrate the gas:

$$\delta = \left(\frac{2}{6wM}\right)^{\frac{1}{2}} > R_c \text{ where } M = \text{permeability} \quad (23a)$$

We can thus summarize the operation (ideal) conditions of the dipolar thruster:

$$BwR_c < \frac{Zkl}{R^2} \text{ (no field ionization)} \quad (24)$$

$$\frac{3}{2}kT < \frac{U_i}{n^2} \text{ (no collision ionization)} \quad (25)$$

$$\frac{3}{2}kT \neq (V_n - V_{n-1}) \text{ (elastic collisions with no absorption)} \quad (26)$$

$$R_c < \delta = \left(\frac{2}{6wM}\right)^{\frac{1}{2}} \text{ (size of channel less than skin depth)} \quad (27)$$

$$P_{dis} = \frac{P_{elh}}{Q} \text{ (electric power discipation is low per unit of thrust)} \quad (28)$$

$$\tau e > \frac{L}{V_g} \text{ (lifetime of Rydberg long enough to accelarated)} \quad (29)$$

Finally, with respect to equation (28), high "Q" circuits are required to reduce electrical losses, which increase the selectivity or narrows the bandwidth of the circuit.

These conditions, as mentioned, may be achievable only in applications where the propellant can be optimumly selected. In other areas, such as coupling with the atmospheric gases, the properties are dictated by the ambient temperature and pressure conditions. This will be more fully appreciated as the following embodiments are described.

NATURE OF EXCITED STATES

A general discussion of excited states in particles such as atoms and molecules and their electric dipolar properties is deemed essential for proper understanding of the present invention. The physical description of the invention has been viewed in a strictly classical manner, i.e., the quantum mechanical aspects of the propulsion concept have not been directly considered. The May, 1981 issue of *Scientific American* contained an article entitled "Highly Excited Atoms" providing a review of excited levels of atoms. An atom or a molecule can be excited by the absorption of a quanta of energy equal to its first transition energy level, around 10 ev. The method of excitation can be from a source of ultraviolet radiation as from a lamp or laser having a photon energy equal to Planck's constant (h) times the frequency, or by the impact of an ion or electron having a translational kinetic energy of approximately 10 ev. A review of electron impact excitation can be found in National Bureau of Standards report NSRDS-NBS 25, dated August, 1968, entitled "Electron Impact Excitation of Atoms." Photons offer the advantage of narrow energy spread and resonant excitation. Electron impact generally gives much less selectivity but creates a more intense population of excited states. In electron impact excitation, intense electron beams or discharges can be obtained and electron impact cross sections tend to be larger than photon cross sections. Both techniques are invisioned as being utilizable with the present invention, depending on the application, one technique may be preferred over another.

Excitation of an isolated molecule may lead to ionization, autoionization, dissociation, predissociation, or reradiation of the excitation energy. Each of the energy excitation processes, can in principle, occur and compete with each other. However, since the rates may differ by many orders of magnitude, usually one process dominates the excitation process. The primary mechanism is currently viewed as being dissociation, especially of oxygen in the air which has the lowest dissociation energy of around 5 ev., nearly half that of nitrogen. At sufficiently high electron impact energies, above 25 ev., the oxygen molecule breaks into two atomic fragments, one being a high Rydberg state and the other a low metastable Rydberg ($3s^5S^o$) at 9.13 ev. Because it is the lowest quintet state, it is metastable with a radiative lifetime of about one millisecond. Rydberg states that have atoms of large principle quantum numbers (n), although not metastable by any selection rules, have long enough lifetimes to be observed in the laboratory. The energy required to remove an electron from a simple atom is given by:

$$E = 13.6/n^2 \text{ eV} \quad (30)$$

The mean value of the orbital radius is

$$r = 0.26(3n^2 - a(l+1))\,A^o \quad (31)$$

where l is the orbital angular momentum integer. For an s electron with $n=20$, this radius is 156 A^o; this radius is huge. The radiative lifetime of a Rydberg state is proportional to n^3 and can therefore reach values between 10 to 100 microseconds for a state with $n=20$, but with an ionization potential of 0.034 eV, it is readily ionized by ambient thermal collisions. Hence, an n value this high represents an upper limit for the present invention which seeks acceleration of an unionized atmospheric gases.

The atoms of a diatomic molecule can rotate about the molecule's center of mass and vibrate along the interatomic axis. The energies of both molecular rotation and of vibration are quantized, and this leads to distinctive molecular rotational and vibrational spectra. The present invention is only concerned with rotation since these represent lower frequencies (RF) whereas vibrational energies usually lie in the infrared. The angular momentum L associated with the molecular rotation of a diatomic molecule is quantized according to the rule:

$$L = J(J+1)\,h \quad (32)$$

where J is the rotational momentum, (I) quantum number with possible values 0, 1, 2, . . . n−1. This quantization implies that the energy of molecular rotation is quantized, and the respective absorption frequency is given by:

$$f = J(J+1)h/2I \quad (33)$$

where (I) is the moment of inertia of the molecule. The moment of inertia is given by:

$$I = m_o n^4 r_o^2 \quad (34)$$

where r_o is the separation distance between the two nuclei of the diatomic molecule. Transitions between the quantized molecular rotational energy states of a polar molecule gives rise to the molecule's pure rotational spectrum. The selection rule governing allowed transitions is J=+−1. The rotational spectrum consists of equally spaced lines typically found in the far infrared and microwave regions of the electromagnetic spectrum for ground states. For excited states, the moment of inertia increases as n^4 and the rotation frequencies may be lowered to radio frequencies. Thus, it is clear that the rotation of water vapor molecules which are polar, to create thrust in the atmosphere in a high frequency Lorentz field, is quantized and selected frequencies are most effective for resonant absorption of energy.

It is also possible to have "superexcited molecules," that is, there is high probability of a molecule receiving energy in excess of its lowest ionization potential without immediate ejection an electron, as such, superexcited molecules form electrically neutral excited molecules possessing energy greater than the ionization potential. Such a superexcited molecule, may, like molecules excited to states below the ionization potential, undergo dissociation to form smaller fragments, one or both of which may be electronically excited.

An electronically excited molecule is thermodynamically unstable, and can lose energy rapidly by several competitive pathways. The actual lifetime of a superexcited molecule depends on its nature, on the complexity of the molecule, and the possible alternative degradation processes. The magnitude of such lifetimes are generally in the very wide range from 10^{10} to 10^{-3} second. One such process is molecular dissociation of the excited state leading to the formation of atoms or smaller molecules, which, in turn, may be excited. In contrast, the most likely processes leading to energy degradation without reaction are radiation conversion (fluorescence), or nonradiative conversion (internal conversion) to the ground state. The latter is generally less probable than internal conversion to the lowest excited state followed by fluorescence to the ground state. Internal conversion is a rapid process (10^{-10} sec), and may include intersystem crossing which involves a change of multiplicity, i.e., transition from a low lying singlet state to a lower lying triplet excited state. Triplet states are potentially very important in the present invention since light emission with a change of multiplicity (phosphoresence) is a slow process ($>10^{-4}$ sec), and the electronic energy is available for comparatively long times to provide longer periods of acceleration. Triplet states may also be formed by direct excitation by slow electrons and in the recombination reaction of a positive ion and electron.

It is clear that fluorescent energy emitted by one molecule could be absorbed by another. However, energy transfer can also occur from excited molecules by a nonradiative resonance process. This is formally equivalent to the emission of a photon by the excited molecule and its absorption by another molecule whose absorption spectrum overlaps the emission spectrum of the emissive molecule. This process is not restricted to situations involving collisions between molecules, but can occur when the distance separating the molecules is less than the wavelength of the emitted photon and can take place efficiently over distances of 50–100 A°.

In the case of collisions between neighboring particles, a pressure dependence of the excitation process involves the following major factors: (1) imprisonment of resonance radiation; and (2) collision transfer of excitation. Reabsorption of photons by atoms in the ground state effectively lengthens the life of the excited state, and spreads the excited state population over a larger volume. The longer effective lifetime of the upper state results in an increased probability for intervention of collisional processes, and for conversion through radiative transitions to lower levels other than the ground state. In collisional transfer, an excited atom is de-excited in a collision with a ground state atom with a transfer of excitation energy and possibly changes in the values of angular momentum and spin associated with the excitation energy. With the addition of gases of different species, "optical pumping" may occur in which the foreign atoms act as buffer atoms such that collisions between the excited atom and the buffer atom will not undo the excited state but because of the shapes of the electron orbits of the two particles, the buffer atoms prevent the magnetic interaction of their electrons. It is by this process that a population of excited or pumped atoms leak back to an unpumped, low ground state. The existence of such processes serves to diminish the excitation power required to accelerate a given amount of gas. The creation of a "population inversion" state is obtained. Thus, laser action may be used for practicing the present invention.

A discussion of an excited state of a single atom versus the ionization state energy will now be considered. The energy required to excite an atom to a given P.Q.N. is given by:

$$U_e = U_i\left(1 - \frac{1}{n^2}\right) \quad (35)$$

Where U_i is the ground state ionization potential. The ratio of the maximum possible dipole moment-to-mass ratio (52) per unit of excitation energy is as follows:

$$\frac{\Omega}{Ue} = \frac{C_1 n^3}{(1 - 1/n^3)} \quad (36)$$

For large values of (n), this ratio increases as a function of n^3. Hence, it appears that the most effective use of the excitation energy occurs at the highest possible (n) value, and just below the threshold of ionization. The absorption frequency, however, becomes smaller as higher states of (n) are reached as shown as follows:

$$\nu = CR\left(\frac{1}{n_l^2} - \frac{1}{n_u^2}\right) \quad (37)$$

where:
n^u = upper Principal Quantum Number
c = velocity of light
n_l = lower

R = Rydberg Constant

At values of (n) greater than about 40, the electronic absorption frequency lies in the microwave region, compared to the ultraviolet region at near ground state values of (n).

In typical laboratory experiments with excited states, high values of (n) are achieved by using a gas laser and a tunable dye laser which provides some control over the frequency range. Thus, the process can be controlled from the ultraviolet to the microwave frequencies. Experiments in the laboratory have been performed with molecular beams of sodium in a high vacuum. A magnetic field is used to extract any ions that are present after excitation. The levels of excitation are then measured by applying a known field ionization voltage between a set of electrodes around the beam. The cutoff voltage will ionize all particles of a specific (n) and higher, providing an ionization current, whose magnitude determines the population of these excited states. Electron impact as well as laser excitation have been used. A discussion of the former, with reference to atmospheric oxygen, can be found in the *Journal of Chemical Physics,* page 3125 by R. Freund, Apr. 1, 1971.

MECHANISMS OF DIPOLAR COUPLING

There are at least five different basic methods of creating dipolar type interactions with an external Lorentz field. These various mechanisms are briefly reviewed as follows:

(a) Electronic dipole—at any instant of time, an electron in its orbit about the nucleus constitutes a dipole, and, as the electron orbits, the system can be viewed as a dipole rotating at the orbital frequency of the electron about the nucleus. This frequency is given by:

$$f=(\tfrac{1}{2}\pi)\,[ke^2/mn^6r_o^3]^{\frac{1}{2}} \quad (38)$$

Hence, for excited states the frequency decreases inversely as the cube of the principle quantum number. For ground states, this frequency lies in the ultraviolet region, and for excited states, the frequencies involved lie well into the microwave region.

(b) Precessing Atomic Orbital Dipoles—The velocity of the electron reaches a low at the perhilion of its electrically stressed orbit which is the region of high space charge concentration. Hence, the flip-flopping of this orbit results in an oscillating dipole that establishes the polarization current density. To create this condition using atmospheric gases, the diatomic molecules must first be dissociated. The dipole moment is the major axis of the elliptical orbit from the more massive ion core to the electron at the perhilion, times the electronic charge.

(c) Precessing Molecular Orbital Dipoles—Here, the particle remains a molecule and the energy of dissociation is avoided. The molecule is excited, and the orbital perhilion is established by the alternating electric field. The dipole motion readily follows the applied electric field. This, together with method (b) represents the most common methods of dipolar coupling for atmospheric gases.

(d) Permanent Assymetric Dipolar Molecules—Some molecules, such as water (H_20), possess a permanent dipole moment (1.85 Debyes) due to the assymetry of the covalent chemical bonds between the constituent atoms. Other common dipoles of this type are NH_3. The rotation of these molecules is quantized, but clearly no energy of excitation is required to attain moderate dipole moments.

(e) Heteropolar Molecules—Some molecules which have ionic bonds, possess permanent electric dipole moments. For example, sodium chloride NaC1, has a dissociation energy of 4.24 eV, and an equilibrium separation distance of 2.36 A°. Since the molecule is held together by ionic binding, the end containing the Na nucleus represents a region of positive electric charge. The end containing the C1 nucleus represents a region of negative charge. Hence, it has a dipole moment of 9.0 Debyes, more than four times larger than the water molecule. Such molecules, while not existing in the atmosphere, could be used in more conventional thruster applications.

Before proceeding with a further discussion of the dipolar force field propulsion system, certain of the well known matter particles which may be useful as a source of neutral particles of matter having selected electric dipole moment or polarizability characteristics with known breakdown characteristics for practicing this invention are set forth hereinbelow. The below list are examples of possible ground state propellants for the dipolar force field propulsion system.

TABLE 2

PROPERTIES OF DIPOLAR SUBSTANCES IN GROUND STATE

Specie	Molecular Weight	Permanent Dipole Moment (D)*	Polarizability (A^3)	Ionization Potential (eV)	Dissassociation Energy (eV)	Boiling Point (°K.)
Helium (He)	4	—	2.5	24.48	—	4.95
Water (H_2O)	18	1.85	18.6	12.6	2.5	
Sodium (Na)	23	23.6		5.138	—	
Ammonia (NH_3)	17	1.47	27.8			239.8
Lithium Floride (LiF)	26	6.33				
Nitrogen (N_2)	28	—	22.1	15.576	9.75	77.35
Oxygen (O_2)	32	2 B.M.**		12.063	5.0	90.18
Hydrogen Cloride (HCl)	36.45	1.08				
Salt (NaCl)	58.45	9.00			4.24	
Xenon (Xe)	131.30	—	27.3	12.127	—	166.05

*D = Debye = 3.3×10^{-30} Coul-meter
**B.M. = Bohr Magnetron

FIG. 5 is a graph showing the polarizability and ionization potential versus the energy level of the atom. However, the field ionization limit of the particle cannot be exceeded, otherwise, ionization will occur, which is undesirable. The ionization potential decreases rapidly as the P.Q.N. increases. Also shown is the electronic absorption frequencies as the P.Q.N. is increased.

The thermal energy of ambient gas molecules is of the order 0.04 eV, hence P.Q.N. of up to 15 to 20 are possible without causing ionization of the excited atom. It is evident that polarizabilities up to 10-24m³ may be possible at ambient temperatures. Quantatively, the following condition must be satisfied:

$$E_f \cong BR_c w < \frac{Ke^2}{R^2} \quad (39)$$

where R is the coil radius.

The coulomb electric field experienced by the electron in its orbit equals the induced electric field at a distance R_c from the coil. In other words, the orbital radius of the Rydberg electron is restricted to the value indicated to prevent ionization of the atom. For the condition just mentioned, (n=20, B=½ Tesla and R=½ meter), the P.Q.N. is limited to 17; hence the Rydberg atom will not ionize for P.Q.N.'s equal to or less than this value. Combining equation (30) with (28) and (24) and solving for the maximum possible atmospheric acceleration "$\ddot{x}$," for a given n. R_c and molecular weight, we obtain:

$$\ddot{x}_{atmos} = \frac{2.2 \times 10^7}{R_c} \text{ M/sec}^2 \quad (40)$$

where R_c is the radius from the center of the coil to the point of interest. This equation gives the approximate limit in particle acceleration that is possible without causing ionization to take place in the atmosphere. The equation further implies that for larger and larger diameter field coils, it is desirable to have lower excited states in order to avoid ionization.

FIG. **6** is a graph showing the possible particle acceleration attainable as a function of the product of the magnetic field and the applied frequency which must be less than the field ionization potential limit. The field ionization limit is not a specific boundary layer condition, but is a range where ionization occurs and is dependent on a number of variables including the strength of the magnetic field, frequency and properties of the dielectric substance. The graph of FIG. **6** is based upon the use of nitrogen (N2), the primary constituent of air for various principle quantum numbers. Also, shown is the acceleration of water vapor molecules which are assumed to remain in a ground state. From the graph in FIG. **6**, particle accelerations of 10^6 to 10^6 m/s² may be possible. These types of accelerations are typical of the gas accelerations found in rocket and jet engine thrust chambers. Hence, this invention has utility in propulsion applications.

LINEAR DIPOLE FIELD ACCELERATOR

Referring now to FIG. **7**, a simple LCR circuit is shown consisting of an electrode pair **140** and **142** having a capacitance C and which contains a polarizable gas **160** therebetween as a dielectric. The capacitor C defined by the electrode pairs **140** and **142** and the polarizable gas **160** as a dielectric is series connected to an inductance coil **146** having an inductance L. The inductance coil **146** provides a crossed magnetic field which crosses and intercepts the electric field extending between the electrodes **140** and **142**. The inductance coil **146** is shown in greater detail in FIG. **9**. The LCR circuit illustrated in FIG. **7** provides an electric and magnetic field which vary, in time, as a cosine and sine function, respectively. The circuit has a resistance R, shown by element **150**, which should be minimized to reduce joule heat losses. The circuit is supplied with electrical power from a voltage source E, shown as **152**, by closing a switch **154**. The gas molecules **160**, which are to be accelerated, are located between electrode **140** and **142** and are excited by a vacuum ultraviolet radiation source **170** having a reflector **172**. The radiation from the radiation source **170** is shown by arrow **174** which is directed into the gas molecules **160**. The gas molecules **160**, when in a ground state, normally have a relative dielectric constant (K_r) near unity. However, when the gas is excited, the dielectric constant, and hence the capacitance, increases significantly as given by the following equation:

$$P = (K_R - 1)\epsilon_0 E \quad (41)$$

where $P = \bar{p}N$, the polarization or average dipole moment per unit volume

For gases excited to PQN=10, the dielectric constant is near 10,000, at 10 KV/M field strength. Therefore, even with small electrodes, significant electrical energy can be stored in the excited gases. In the perferred embodiment, the electrodes **140** and **142** are sized to store the energy cycled between the coil **146** and capacitor C having the excited gas as a dielectric. The entire circuit is tuned for operation at substantially the resonant frequency. This configuration establishes the requisite spatial and time force field region to generate a dipolar propulsive effect on the gas.

Referring now to FIG. **8**, a linear accelerator **200** is shown using the construction and elements of the simple LCR circuit shown in FIGS. **7**, **8** and **9**. The linear accelerator **200** consists of a number of electroconductive plates **202** which act capacitively and are arranged to form two sides of a linear rectangular channel **206**. The other two sides of the linear rectangular channel **206** are the pole faces **212** and **214** of a series of U-shaped electromagnets which are arranged along the channel length. The electromagnets have energizing coils **216** situated externally to the linear rectangular channel **206**. The alternating current power source is applied to a conductor **220** across plate **202**, which plates are connected in series with the windings **216** of the coils **212**. The other side of the windings **216** of coil **212** is connected to conductor **222** which, in turn, is connected across the alternating current source. The windings **216** of coil **212** and the capacitive electrode **202** are electrically connected in series as shown in FIG. **10**. The connector **220** and **222** are responsive to an alternating current power source to provide a crossed electric and magnetic fields across the electrodes **202** and the windings **216** which vary as a cosine and sine function, respectively. Each stage of the elements which define the linear rectangular channel **206** are connected in parallel to each other to reduce the equivalent reactance to permit high frequency (HF) operation. The circuits are driven from an external high frequency power source which is applied via a control means **250** such that the frequency of the alternating current power supply is adjusted to operate at the resonant frequency of the circuit.

The gas molecules, which is to form the dielectric gas to be accelerated, is stored in a cryogenic Dewar of gas which maintains the dielectric gas at a extremely cold temperature. The cryogenic Dewar of gas is illustrated as **252** in FIG. **10**. The gas is stored in a pressurized vessel **254**. The dielectric gas passes from the pressurized container **254** through a regulator **256** to a control valve **258**. The control valve is operatively coupled to a "U" shaped cooling member **260** which passes along each stage of the linear accelerator and which is located under each of the windings **216** of the coils **212** to provide cooling of the magnets **212** to increase the conductivity thereof. The "U" cooling core **260** has its other end terminating in a flowmeter **262**. The other side of the flowmeter **262** is adapted to feed the gas to a plenum **280** which, in turn, distributes the extremely cold dielectric gas into the linear rectangular channel **206**. The flow meter **262** is utilized to control the flow of the dielectric gas into the longitudinal rectangular channel **206**. Any suitable cold gas may be used, such as an inert gas which is chemically inert and, as such, avoids causing corrosion to the electrodes. Preferably, the gas pressure in the longitudinal rectangular channel **206** is as low as possible to reduce collisional quenching of the gas.

The dielectric gas located in the longitudinal rectan gular chamber **206** is excited by an external excitatioı source such as for example a beam of vacuum ultravio let radiation from a source such as a lasar **270**. The lasa **270** is bounded at one end of the longitudinal rectangu lar channel **206** and is positioned with respect theret such that the lasar beam tranverses the entire insid length of the longitudinal rectangular channel **206** s that the dielectric gas contained therein is constantl exposed to this ultraviolet (UV) radiation. The dielec tric gas is excited by the UV radiation and is strongl coupled to the alternating cross field from the electro magnets **212** and accellerated as described hereinbefore The electric field utilized in this embodiment appear across the capacitors defined by the plates **202** havin the dielectric gas therebetween. The magnitude of th electric field utilized in this embodiment is determine by the voltage that appears across the capacitors de fined by the plates **202** along with the dielectric ga stored therebetween.

As stated hereinabove, the dielectric gas is preferabl supplied from a Dewar 252 which preserves the fluid a a cryogenic fluid (such as helium at 4.4 degree K). Th dielectric gas supplied is preferably as cold as possibl to reduce the collisional thermoquenching of the ex cited states which is controlled by the following for mula:

$$3/2kt = ui/n^2 \quad (42$$

With a cryogenic dielectric gas, the P.Q.N.s over 10 might be possible. Thus, such a dielectric gas may hav a very large electric dipole moment in high partic accelerations at low field frequency products. The pos sible P.Q.N. is determined by the following equation:

$$P.Q.N. = \left[\frac{2U_i}{3kT} \right]^{\frac{1}{2}} \quad (43$$

The electric field utilized inthis embodiment is tha across the capacitors ("Q" times the supply voltage V_s as it alternates its stored energy with the magnet coil according to:

$$\tfrac{1}{2}cu^2 = \tfrac{1}{2}Li^2$$

where: C is the capacitance
V is the voltage across electrodes
L is the coil inductance
i is the coil current
Hence,

$$V = [L/C]^{\frac{1}{2}}\, i \text{ and } E_c = \frac{V}{d} \quad (44)$$

where d = electrode gap

The induced electric field (E_i) due to the time varying magnetic field which has a direction at right angles to the magnetic field is not utilized. This capacitive electric field is more useful at lower frequencies when:

$$E_c >> E_i \quad (45)$$

Whereas E_i is useful at higher frequencies and magnetic fields as described in other embodiments later.

FIG. **9** illustrates one embodiment of a linear dipole field accelerator which can be used for practice in the invention. The accelerator includes electrodes **312** and **314** which are adapted to establish electromotive lines of force thereacross to establish electric field as illustrated by the arrows **318**. Electrode **312** is adapted to be connected via conductor **320** to an alternating electric field source. The other electrode **314** is connected via conductor **330** to windings **332** of a coil **334**. The other end of the windings **332** of coil **334** terminates in a conductor **340** which is adapted to be connected across the other side of the alternating electric field source. A highly permeable magnetic conductive member, **360**, (such as ferrite) generates the necessary magnetic lines of force which are shown in FIG. **9** by arrows **362**. The magnetic lines of force extend in a direction which is at a predetermined angle to selected or first direction of the electromotive lines of force and the magnetic lines of force **362** cross and intercept the electromotive lines of force **318** at a predetermined location to define a spatial force field region which is located between the electrodes **212** and **214**. The alternating electric source which is applied between conductors **320** and **340** generate both electromotive and magnetic lines of force as a function of the magnitude of the alternating source which varies as a function of a selected frequency. Thus, since the electroplates **312** and **314** are connected in series with the coil **334**, the rate of frequency change of the alternating source will determine the frequency of the electric field applied across the electrodes **312** and **314** and the frequency of the alternating magnetic field developed by windings **332** and applied via the magnetic coupler **360** across the spatial force field region.

FIG. **10** is the schematic diagram which shows a schematic for a multistage linear dipolar field accelerator having five stages shown by stages **364** through **372**, inclusive. The stages are all connected in parallel across a pair of conductors **374** and **376** which energize from a source **380**. Stage **372** illustrates a capacitor **386** which is formed of electrode plates in a manner similar to that described with respect to FIG. **7** wherein the polarizable gas molecules are located between the plates of the capacitors and wherein the gas molecules has a predetermined dielectric characteristic. The inductor, shown by inductor **388**, is formed of high-flux density magnetic coil.

FIG. **11** illustrates the potential specific impulses that appear to be possible, using the principles of this invention, as the operating parameter of the device. The operating parameters are the product of the channel length and the frequency and magnetic field. The graph of FIG. **11** shows the Isp for various dipole moment-to-mass ratios, such as water, and excited gases. The later can have ratios equal to or greater than unity, if the gas is properly cooled to minimize thermal quenching.

ELECTRICAL POWER REQUIREMENTS

The generation of thrust utilizing the principles disclosed in this Application requires the absorbtion of power by the dipoles in the gas. Generally, this energy absorption can be grouped into five different categories:

(1) Excitation Energy—energy in the form of a quanta of photon (hv) or electron impact energy is absorbed to create an excited state having a high polarizability or induced dipole moment.

(2) Orientational Energy—thermal molecular collisions in the gas tend to disorient the dipoles in the external applied electric field; consequently a restoring torque equal to pXE must be applied.

(3) Polarizability Energy—once the atom is excited, the electronic cloud must be stressed or distorted to create an induced dipole, with energy given by $\frac{1}{2}\alpha E^2$.

(4) Rotational Energy—finally, the dipole must be rotated by an alternating electric field, and since it has a finite moment of inertia, it has a rotational energy $\frac{1}{2}$ I W^2 which must be maintained regardless of molecular collisions.

(5) Translational Energy—the kinetic energy of the particles ($\frac{1}{2}$ mV^2).

In general, most of these energies are very small compared to the excitation energy required, which energy cannot exceed the ionization potential of the gas, around 14–15 ev for atmospheric gases. In addition to these energies, which are absorbed by the dipoles, there are various losses that the system will incur:

(1) Radiation Losses: Once an excited state has been accelerated and is quenched or deactivated, it will flourese or emit a photon of radiation which may be aborbed by another neighboring atom or lost to the system. If the emitted radiation excites another atom, then this improves efficiency as the energy of excitation is reused. Finally, the coil itself is a RF antenna that broadcasts radio frequency energy which can be reduced by correct design or shielding.

(2) Thermal Losses: The coil has a resistance which generates a joule heating loss (I^2R) which must be minimized or reduced by cooling to prevent overheating the coil. The use of cryogenic cooling or superconductivity is exploited in this respect.

Further, dielectric losses in dielectric gas are reduced. A circuit diagram of the power source and electrically coupled load was sown in FIGS. 7 and 10. In order to achieve sufficient thrust density at lower frequencies, high magnetic fields in the vicinity of 0.1 to 1 Teslas are desired. The stored magnetic field energy in the working volume times the cycle frequency represents the circulating electrical power. The actual power discipation is the circulating power divided by the circuit "Q", or figure of merit which is the ratio of inductive reactance to the resistance. The ratio of the body force to the body power discipation thus simplifies to:

$$F_b/P_{dis}=K_1\lambda_e\omega^2/R \quad (46)$$

Thus, for a given condition of excited gas or electric susceptability, the ratio of the frequency to resistance (Q) should be optimized. The coils shown in FIG. 10 thus consist of elements of large cross-sectional area with minimal length and are cooled to very low temperatures to minimize the resistivity. For example, in a rocket driven MHD power generator, the liquid hydrogen (−400° F.) for the fuel can be circulated through the coil before combustion takes place.

The present invention has utility as a new and innovative propulsion system in which the thrust-to-power-ratio is potentially very high compared to conventional systems. The thrust-to-power ratio (γ) is given by:

$$\gamma = \frac{\dot{m}V_g}{U_e\dot{N}_e + \frac{1}{2}\dot{m}V_g^2} \quad (47)$$

where:
$\dot{m}$ is the total mass flow rate (kg/s)
U_e is the excitation (photon) energy absorbed
$\dot{N}_e$ is the number flow rate of excited particles
V_g is the net change in velocity of the gas

If (β) equals the population fraction of excited states in the total gas flow, this equation becomes:

$$\gamma = \frac{2v_g}{\frac{2U_e\beta}{M_o} + V_2} \quad (48)$$

Where M_o is the mass of the atom or molecule. Differenting this equation and setting equal to zero, we find the optimum velocity for maximum (γ):

$$V_{optimum} = \left[\frac{2\ U_e\beta}{m_o}\right]^{\frac{1}{2}} \quad (49)$$

As an illustrative example, assume an excitation photon energy of 10 ev and a population fraction of 1% or 0.01 using diatomic nitrogen with a mass $m_o=28\cdot1.67\times10^{-27}$Kg, the velocity is 840 m/s and the power/thrust ratio is about 800 watts/newton. This compares to the performance of the SSME rocket engine on the space shuttle which requires 4540 watts/-newton of thrust. Even better performance may be possible depending upon the number of rebounding collisions and collision cross-section of the excited atoms, which are generally much larger than ground state atoms.

FIG. 12 is graph showing the power/thrust ratio for atomic hydrogen gas assuming no photon recycling the ratio decreases inversly with the square root of the molecular weight, thus Xenon gas would have a power thrust ratio more than 10 times smaller than hydrogen. Moreover, the ionization limit is moved further up so that higher induced electric fields are possible without field ionization. In fact, the field ionization limit (E_i) increases as follows:

$$E_i = \frac{ZKe}{R^2} \quad (50)$$

It increases with atomic number for an atom of given radius (R). As illustrated in connection with FIG. 7, no effort was made to capture "lost radiation." It was simply assumed that the gas completely absorbed the UV radiation as it traversed the length of the acceleration channel 306.

As shown in FIG. 13, the input gas 400 located between plates 402 absorbs the photons from an excitation source 406. The gas 400 is excited for a lifetime τ_e and then is de-excited. Meanwhile, the gas has traveled a length ($V_g\tau_e$), where V_g is the gas velocity. The de-excitation involves the emission of a photon, with a frequency generally less than the original frequency, but still greater than that required for the first transition state above ground and thus, can be usefully "recaptured." Thus, FIG. **13** shows the emitted photon **412** being reflected (arrow **414**) and returned upstream to the source gas where reaborption takes place. In practice, the input radiation may be introduced at right angles to the gas flow, and bounce repeatedly off the walls of the channel, which are approximately made into reflecting surfaces.

In the embodiment illustrated in FIGS. **14** and **15**, the feature of reflecting trapped radiation in a optical cavity is utilized. In this embodiment, a torroidal coil is used to establish the alternating magnetic field between a cylindrical capacitive electrode arrangement. The torroid coil has the advantage of having no external field (for the ideal case) and, consequently radiation losses are minimal.

Referring to FIGS. **14** and **15** therein is shown this particular embodiment consisting of flat rectangular plate conducting elements **450** arranged around a pair of cylindrical electrodes **452** and **454**. The plate conducting elements **450** are insulated from the cylindrical electrodes **452** and are held to the core conductor by a collet arrangement with spindle chuck assembly **458** which locks them into position. A source of UV radiation **470** enters through holes **472** such as from an exciter laser source **474**. The UV beam is tilted slightly off a radius vector to allow the beam **470** to be reflected off the inner reflective surfaces of the cylindrical electrodes which also act as an optical cavity to trap the radiation. The air or dielectric gas, shown by arrow **420**, enters from the left through the conducting elements **450** and is immediately excited by the radiation and electromagnetically accelerated. As mentioned earlier, an alternate method of excitation involves electric discharges which should also be considered in this application. In this embodiment, the plate conducting element **450** are connected in series with the cylindrical electrodes **452** which define the capacitors.

METHODS OF EXCITATION

As mentioned previously, there are two basic methods of excitation involing (1) electron impact and (2) radiative or photon interaction. Each approach has its advantages and disadvantages and which one or both should be utilized depends upon the application.

Methods employing electron impact are:

Electron Beam Excitation—in this a cathode is heated in an evacuated chamber and when a voltage is applied, electrons are emitted which can be focused and directed into the gas. The beam tends to be rapidly attentuated in the atomosphere and diverges with distance due to mutual repulsion between the electrons. This technique might be used directly in propulsion applications of small dimensions, comparable to the attentuation path length.

High Voltage AC or DC Electric Discharge—this technique is perhaps the easiest to implement, can be lightweight and may provide good efficiency. Using the AC approach, the voltage may be readily stepped up to high voltages, e.g., by using a Tesla coil. The breakdown voltage causes ionization to take place, and the ions and electrons, in turn are accelerated by the field to impact ground state atoms to cause excitation. The DC approach is more complex insomuch as HV rectifiers are required, and it's not clear what is gained by doing it this way.

Radio Frequency or Microwave Discharge—in this technique, a high power microwave is applied to the gas, which, by heating the gas leads to thermal ionization and excitation. Once some ions are generated, they are further accelerated by the fields to cause more excitation and ionization. This method may not be the most efficient since thermalization and ionization may dominate the process with only incidental excitation to take place. However, if it can be made efficient, it promises to be operative over larger volumes. If done at lower megahertz frequencies, the field coil of the propulsion system itself may be used to achieve self-excitation.

Methods employing photon interactions are:

Flashtube or Flashlamp Excitation—in this technique a Xenon flashlamp is fired with a high voltage pulse which emits a spectrum of light of varied frequencies. The efficiency is low, and, moreover, there exist many frequencies not useful, i.e., that do not conform to an energy transition in the atom or molecule to be excited. Even so, the radiation can be directed, reflected in an optical cavity, and can penetrate the gas over large distances.

Laser Beam Excitation—This technique offers the advantage of a single monochromatic beam of intense, coherent electromagnetic radiation. A wide variety of types of lasers exist, e.g., water vapor lasers, nitrogen (pulsed) are rare gas excimer lasers than emit in the ultraviolet range, with photon energies that overlap the transitions in the atoms to be excited (about 10 eV), or around 1300 A wavelength. The required resonance transition levels may be 1300 A wavelength. The required resonance transition levels may be easily excited by a low pressure electrodeless discharge sustained in a microwave generator, and the resultant photons transmitted into the reaction chamber or channel through lithium floride sapphire, or calcium flouride windows. The exciting wavelengths provided by such sources include xenon (1295 A at 9.6 eV); argon (1048 A at 11.8 eV and 1067 A at 11.6 eV). When the photon energy is less than the ionization potential, the invention can function in the absence of ionization. Above the ionization potential, superexcited molecules may occur, with the added possibility of ionization. The efficiency of these lasers is generally only a few percent, but efficiencies of up to 10% for chemical lasers has been reported.

Synchrotron Radiation Sources—These utilize the acceleration of relativistic electron beams to produce radiation. The possibility of FEL's, or Free Electron Lasers may mean electron beams interacting with "wigler" magnetic fields to generate coherent radiation, may provide up to 50% efficiency.

FIG. **16** illustrates the mechanical efficiency of energy conversion into vehicle kinetic energy by reacting against gas of large mass via a force field extending over space.

FIG. **17** illustrates the range of required force field (R_e) plotted as a function of decreasing medium density. As illustrated in FIG. **17**, the medium density decreases, that is the density of the atmosphere from sea level to interstellar space. Thus, the range of the force field (R_e) increases inversely to the medium density. A family of curves are plotted for various mass ratios (M.R.).

FIG. **18** is another embodiment showing the construction of a wing structure which functions as a capacitor. The wing structure includes an exterior metal surface **500** having a plurality of cell members joined by conductors **502** which are wound in a circular pattern therearound. The center of the wing member is insulated with an insulating material shown as **504**. The wings have the voltage applied thereto to generate an E field which extends perpendicular from the surface thereof illustrated by arrows **506**. The magnetic lines of force illustrated by arrows **508** establish a B field which crosses with the E field as illustrated in FIG. **18**.

FIG. **19** shows the construction of the wing member illustrated in FIG. **18** in a top view. The wing is divided into a plurality of sectors designated as **512** through **526**. Sector **520** has a first electrode **530** which is electrically connected to a spiral shape conductive member **532** which extends through the various sectors as illustrated by the dash line in FIG. **19**. An electrode **534** that is located in sector **518** is adapted to be connected to a spiral connector **532**. In a similar manner, sector **516** has electrode **536** and sector **514** has electrode **538** which is, likewise, electrically connected to the spiral conductor **532**.

Also, sectors **522**, **524**, **526** and **512** have wiping contacts **540**, **542**, **544** and **546**, respectively, extending from the opposite side thereof and toward the center opening defined by the sectors. The wiping contact **540**, **542**, **544** and **546** are adapted to be contacted by a wiping member **550** which is in turn connected via a bus connector **552** to the spiral winding **532**. The wiping member **550** functions as a switch which is adapted to connect any one of the sectors **522**, **524**, **526** or **512** to the electrical sprial connector **532**. Any one or more of the other sectors **514**, **516**, **518** and **520** can be electrically connected to a source by appropriate section of the electrodes **538**, **536**, **534** and **530**, respectively. The B field is generated by appropriate magnetic means located in a central opening **554** and the B field is shown emanating from the central core **554** by means of the vector dots **556**.

FIG. **20** is a schematic diagram illustrating the electrical connections of the conductive and capacive elements illustrated in FIG. **19**. The corresponding plates forming each side of the capacitors are illustrated by the same numbers in FIG. **20** as are pictorially represented in FIG. **19**. For example, sector plates **522** and **524**, which are physically located in opposite positions to each other in the sector circle, define one capacitor. The switching member **550** is illustrated as being equal to any one or more of the capacitive elements so as to control the thrust direction. By appropriate switching of the wiping number **550**, the thrust can be controlled so that the spacecraft moves ahead along the arrow designated as "N" in FIG. **19**, or in an alternate direction indicated by the term "NE" in "NW". In FIG. **20**, a reversing switch shown as element **560** can be utilized to reverse thrust of the aerospace vehicle illustrated by FIG. **19**. The alternating source and the inductor coupling means are illustrated generally as **562**.

FIGS. **21**, **22** and **23** show an embodiment of the present invention in which a single wing disc shaped vehicle is presented. This vehicle has the feature of VTOL takeoff as well as conventional horizontal aerodynamic takeoff. The wing electrodes **600** are so contoured that they act to provide aerodynamic lift, as seen more clearly in the side view of FIG. **22**. Fhe craft is powered by a rotating bed nuclear reactor **602** driven by a motor **604**, which is selected to be capable of generating 1 thermal gigawatt of power in a relatively small (nearly 1 ton) device. The air, shown by arrows **610**, enters the inlet **612** and is heated to about 3000° F. by the rotating nuclear bed reactor **602**. The hot working gas turbine engine **616**, which, in turn, drives a high frequency generator (**620**) via a clutch plate **622**. Other methods, such as magnetohydrodymic (MHD) power generation are also possible as described with respect to FIG. **50**. The high frequency generator **620** power output is inductively coupled by a transformer to a wing coil (**632**) via primary winding **624**. The wing coil conductor elements **626** also act as airfoil shaped struts that form the rigid structure of the wing. This is to reduce weight as well as spread the force field over a larger area and couple with more gas. This is shown more clearly in FIG. **27** which shows a frontal sectional view of the vehicle. The wing conductor struts **626** are connected to a common rim bus-bar **630** that ties the coil to the wing electrodes as shown in the electrical schematic as shown in FIG. **23**, formed into struts **632** separated by insulation **634** so as to reduce eddy current losses induced by the alternating field from the wing coil below the electrodes. Flashtubes **640** enclosed in reflectors **642** are provided along the fuselage or hull of the vehicle above and below the wing. As illustrated in FIG. **24***a*, an internal capacitor **642** is provided for internal tuning for "vertical thrust." As described earlier, other methods of excitation are possible, but flashtubes are easiest to illustrate, although low in efficiency. FIG. **23** shows a frontal view of the vehicle with air intake **612**. This view more clearly shows the radiation field emanating from the flashtubes, which fall in 4–90 egree sectors or quadrants.

Referring to FIG. **24***a* and **24***b*, simplified electrical schematic are provided. The circuit of FIG. **24***a* contains the wing coil inductance Lw, shown as **650** and two capacitance electrodes, one internal to the vehicle C_i, shown as **630** and the other the exterior wing electrode capacitance, C_e, shown as **652**. A switch **654** is provided to permit tuning the coil **650** by either one of these capacitors. If the wing electrode pair A and B is switched in, the exterior electric field produced interacts with the magnetic field to generate a horizonial thrust component. Whereas, if only the internal capacitor **652** is switched in, the induced electric field from the time varying magnetic field of the coil generates a vertical thrust component. If the relative two capacitors can be intermediately contacted in a manner familiar to those skilled in the art, any thrust component intermediate to the horizontal and vertical can be generated for directional control as required.

FIG. **24***b* is a schematic diagram of the coils (**660**) forming the inductor on wing electrode **600** shown in FIG. **24***a* for generating the magnetic field for the spacecraft.

FIG. **25** illustrates the mechanism of momentum exchange between an excited molecule (electronic) and a field of ground state molecules. By this process of rebounding collisions, additional impulse is provided with little added energy. The process begins by the absorption of a photon of energy by the particle which becomes more electromagnetic responsive and is thereby accelerated downward by the high frequency Lorentz force field. Only the momentum component in the Z-direction (thrust) is shown. The excited molecule or atom has an increased collision cross section which effectively increases the collision frequency. Because the mass of the excited particle is equal to the ground state molecular mass, the momenta is simply exchanged upon collision. If the energy of collision does not correspond to a transitional energy gap (rotational, vibrational or electronic), of either molecule, the collision will be perfectly elastic. The graph shows the reference line (horizonal) translated back to the top of the graph after each collision to keep the motion depicted within the boundaries of the graph. During the collision process, the excited molecule may gradually decay with the emission of a photon. As a consequence, the dipole moment may decrease, with a resultant diminishment in the momentum imparted by the force field. However, the radiation emitted may be absorbed by another adjacent excited or ground state molecule, so that the photon energy is repeatedly utilized until the gas eventually thermalizes (by which time the gas has already been fully accelerated).

The attainment of high thrust for the least amount of power requires few as possible excited states with large collision cross sections transferring their momentum to the greatest number of ground state atoms. Thus, the graph of FIG. **25** shows the momentum or impulse exchanged versus the number of collisions experienced by an excited atom before it is quenched. The rebounded excited atom is turned each time by the force field and collides with additional ground states. If the dissipation of energy is minimal, the excited state can undergo many collisions in this way before it is extinguished by quenching or radiative decay (deactivation). For example, the sea level collision frequency is 10^9 Hz in air; if the lifetime is 10 microseconds, the total number of collisions possible is $10^9 \times 1.0 \times 10^{--5} = 10^4$ collisions. Accordingly, the momentum induced in the excited Rydberg particle is transferred to thousands of ground state atoms. In this arrangement a very low B field is possible while securing high performance. This is further realized when one considers the large collision cross section of an excited particle relative to a ground state atom; it can be millions of times larger since it increases with the fourth power of the P.Q.N., (n^4). The effect of the larger collision cross section is to increase the collision frequency, which is directly proportional to the cross section.

The propulsion efficiency (thrust power divided by rate of propellant energy release) shown in the graph of FIG. **26** is for three different classes of propulsion systems: rockets, conventional air breathing ramjets or jet engines and a force field propelled system as disclosed herein. The propulsion efficiency equations for the rocket and airbreather respectively are:

$$\eta_{rocket} = \frac{2\nu}{1 + \nu^2} \quad (51)$$

$$\eta_{air\ breathen} = \frac{1}{\frac{\beta}{2\nu} + 1} \quad (52)$$

where

ν=ratio of Vehicle Velocity to exhaust and

β=ratio of delta velocity of air to exhaust of a rocket.

The present force field propulsion system is an air breather in which very low delta velocities are possible due to the interaction with a very large volume or mass of air with dimensions comparable to the size of the vehicle itself. An external force field arrangement could be used in the arrangement. As illustrated in FIG. **24**, rockets gradually reach peak propulsion efficiency as their vehicle velocity approaches their exhaust velocity. Thereafter, the efficiency thereof gradually tapers off. In a ramjet or jet engine, the efficiency gradually increases in a slow and steady fashion. However, when the spacecraft reaches high altitude where the atmosphere density becomes too rarified, the jet engine must be shut down. This occurs at about 100,000 feet. In a force field, air breathing system, operating at low delta velocities over large volumes, the efficiency more rapidly increases at lower vehicle velocities and maintains nearly 90%+ efficiency as velocity increases. Such engines can continue operation at nearly twice the altitude of conventional air breathing engines, with electrical power being supplied by some internal primary energy source such as a nuclear reactor.

FIG. **27** illustrates the possible body force plotted as a function of magnetic field frequency in Tesla-Hertz for a fully excited nitrogen gas at the quantum level of n =10. he plot is for different altitudes commencing at sea level, 50 kilometers and 100 kilometers. When the magnetic field frequency approaches approximately 10^8, field ionization limit is reached which is illustrated by dashed line **680**. The field ionization limit is that point where the gas commences to ionize which reduces the efficiency of the dipolar force field propulsion system.

FIG. **28** is a plot of the body force for various levels or fractions of excitation plotted as a function of the magnetic field frequency in Tesla-Hertz for gas excited at the quantum level of n=10. When the product of the magnetic field times the frequency approaches 10^8 and the population fraction of excited states approaches 100%, the body force is extremely high. A field ionization limit occurs at about 10^8 Tesla Hertz as is illustrated by dashed line **682** in FIG. **28**.

FIGS. **29, 30, 31** and **32** show the construction details of a spacecraft generally referred to as a "X-wing" aerospace vehicle which is adapted to utilize the teachings of the present invention.

The spacecraft includes a lower set of wings **700** and an upper set of wings **702**. If desired, the angle between the upper and lower wings can be variable for efficiency optimization purposes. The aircraft utilizes a verticle tail **204** and horizontal stabilizing fins **706**. A source of electromagnetic radiation, such as an elongated flash tube **708** is located on the lower wing **700** and positioned to direct the electromagnetic radiation generated thereby toward the undersurface of the upper wing **702**. The upper wing **702** includes prismal reflecting member **716** which are adapted to reflect the ultraviolet radiation designated by arrow **714** between the upper and lower reflective surfaces of the wing **700** and **702**. The final radiation is return reflected by reflector **718** located at the extremity of the upper wing. At the terminus of each wing is located a pressurized liquid hydrogen storage tank **720**. The front plan view of FIG. **29** shows that the aerospace vehicle includes air intakes **706**, has a fuselage **722**, landing wheels **726** and, if desired, auxiliary airbreathing jet engines **724**.

The details of the construction of the wings is illustrated in greater detail in FIGS. **30**, **31** and **32**. The inductive coils are formed by strut numbers **730** which are adapted to be a plurality of spaced aligned members and which are adapted to carry a current therein as illustrated by the current flow arrow. The strut members **730** are covered by a conductive surface **734** which function as the conductor plates for confining the dielectric gas therebetween. In the preferred embodiment, the main power plant for generating the alternating current power may be a rotating nuclear bed reactor which is similar to that illustrated in connection with FIG. **21**. The blades of the turbine are illustrated as **740**, the high frequency generator illustrated at **744** which is coupled to the rotating nuclear bed reactor by the clutch **742**. The strut number **746** of wing **702** function as part of the secondary winding of the transformer type coupling member which is operatively coupled to the high frequency generator **744**.

FIG. **31** illustrates in greater detail the construction of the upper and lower wings and the means for generating the electromagnetic field and the electromotive lines of force to establish the E field. The excitation source **708** generates the electromagnetic radiation **714** which is reflected from the optical surfaces of the wing **702** which functions to excite the atoms of nitrogen gas in the atmosphere to a higher quantum level. The gases are confined between the upper wing **702** and the lower wing **700** which establishes the E field shown by lines **752** which pass between the wings and from the pointed ends of members **716** and the B field which emanates from the fuselage, as line **754**. Thus, the area between the wing **700** and **702** function as a spatial force field region which has the excited nitrogen gas particles located therebetween and which, in the presence of the crossed magnetic field and electric field, cause the dipoles thereof to rotate and cause the reactive thrust.

The details of the wing construction disclose that the upper surface of wing **700** is conductive while the lower surface **756** is an insulator. Internal struts **730** are insulated from the upper surface of wing **700** by insulator spacers **750**. Also, each of the struts **730** contain passageways **758** which is adapted to permit hydrogen liquid **760** to pass therebetween. The hydrogen gas acts as a coolant in addition to being used as a fuel and can be utilized to cool the superconducting magnets which generate the magnetic field indicated by arrows **754**.

FIG. **32** illustrates, by means of a cross section, the relationship between the upper wing **702** and the lower wing **700** and the specific construction of the various wing struts. The upper wing **702** is insulated from a center support **762** by an insulator **764**.

In a similar manner, the lower wing **700** has the center strut **720** insulated from the conductive upper surface **774** by means of insulators **750**. Wing struts **730** have the lower outer surface which is formed of insulating materials **756** affixed thereto. The airflow between the wings is illustrated by arrows **778**. The direction of the B field is illustrated by vectors **754** which are extending outward from the fuselage toward the end of the wings. The electromotive lines of force of the electric field as shown by lines **752** and extend between the lower wing **700** and the upper wing **702**.

FIG. **33** is a schematic representation of the inductance coils and electrodes forming the same **786** which are located in each of the wings. The inductors are driven from a high frequency alternating current source through a transformer coupler illustrated as **788**.

The power source which is adapted for use in the "X-wing" aircraft is illustrated in FIG. **34**. In operation, a power source, such as a turbine **790**, drives a high frequency generator at the selected frequency. The high frequency output is coupled through a transformer coupler **794** to the wing and to the inductors **796** which are connected in series with the capacitors **800** formed between the upper and lower wings.

The embodiment of the invention shown in FIGS. **35** and **36** utilizes the inductive electric field due to the motional magnetic field as given by Faraday's Law:

$$\int E \cdot dl = -(d0/dt) \tag{53}$$

For a solenoidal coil, the Azimutha L electric field produced is given by the following equation:

$$E = -\frac{1}{2} R_c \frac{dB}{dt} \tag{54}$$

or

$$E =; \frac{1}{2} R_c W B_0 \cos wt$$

Combining with equations (1), (6) and (54), we obtain:

$$F_b = ; \tfrac{1}{2} t_0 \lambda_e R_c B_0^2 w^2 \sin w^2 t \} \lambda_e = N\alpha \tag{55}$$

the electric susceptability

which is the body force produced in a dielectric gas subjected to an inductive high frequency magnetic field. The force increases with the square of the magnetic field and frequency. An upper limit is reached when the induced electric field becomes so intense that electrical breakdown and ionization of the gas takes place. The invention is preferably operated at such a frequency and magnetic field condition so as to avoid ionization and the problems which would thereby ensue. It should be pointed out that in the presence of a transverse magnetic field, the breakdown voltage of a gas is increased significantly. FIG. **21** is a graph of the potential body force established in the atmosphere for various altitudes for an assumed excited state gas having a P.Q.N. of 10. The upper limit where ionzation will approximately start to take place is also shown in this Figure.

For sea level, the field ionization limit is reached where the product of the magnetic field times the frequency reaches 10^7–10^8 volts per meter. At higher altitudes, this number decreases as the ambient conductivity increases. Even so, body forces of 10^3–10^4 NT/m^3 are possible at megacycle frequency at sea level. This has been discussed in detail with respect to FIGS. **27** and **28** hereinbefore.

As shown in FIG. **35** and **36**, the magnetic field is generated inside of a conical shaped spiral coil **810** consisting of a number of turns each parallel connected to minimize the inductance to permit resonant operation in a series tuned circuit at megacycle frequencies. The coil is preferably made of lightweight material such as aluminum alloy and housed in a structure **812** designed to handle the mechanical stress of the magnetic field pressures. The coil is cryogenically cooled via input flanges **820**, and hollow conductors, with an exit plenum **822**. If the coolant is water (which is a dipole), it may be injected as a fine spray via conduit **224** into the acceleration cavity **826** to enhance the thrust and reduce the levels of excitation required. The water vapor may also be the combustion products of a liquid hydrogen and oxygen rocket driven MHD generator.

The dipolar propulsion unit has an entrance or intake manifold or shroud canapy **830** through which the working fluid such as air enters and is directed into the Lorentz propulsion chamber cavity. The coil elements **810** consist of flat strips through which air is free to pass and are held in rigid position by the insulator attenuator fins **814** which also act to attentuate the exterior unwanted upstream electric field. A source of ultraviolet excitation radiation such as an excimer laser **832** is provided which directs its beam into an optical cavity **834** consisting of a reflecting fresnel surfaces **880** on the conductor strips which bounces the beam **850** back and forth between the surfaces hundreds of times to increase the absorption pathlength and permit more efficient utilization of the radiant energy. The wavelength of the excitation source is choosen such that the photon energy (hv) lies just below the ionization potential of the atoms of the gas, e.g. 1300 Angstroms wavelength. The radiation is thus readily absorbed by the gas and converts the ground state neutral atoms or molecules into highly excited Rydberg or metastable states that more readily electromagnetically coupled to the high frequency magnetic field. It is particularly important that this excitation source have a high energy transformation efficiency to minimize overall power consumption. Electron impact may also be used as described earlier, using the output, e.g. of a Tesla coil. For low velocity, high volume applications, only a small fraction of the total ground state number flowrate into the unit need be converted into an excited state. Additionally, electrodes may be added to provide directional control of forces.

In summary, the device operates as follows: Air enters the upstream entrance **830**. No electric field is experienced because the insulating fins or struts attenuate the field on the upstream side. The air moves through the passages between the conducting strips **810** and is immediately excited at the same time high frequency polarization currents are induced in the gas. The gas is thereby accelerated to a moderate exit velocity at a very large mass flow rate. The residence time during which the gas is accelerated is at least equal to or less than the lifetime of the excited states in the gas, such as metastable oxygen. An alternate method of excitation involves the applications of a attentuating current high voltage to ionize some of the air and excite atoms by electron impact.

The dipolar force field propulsion system has wide application, particularly as a propulsion means for aerospace vehicles such as spacecraft. The aerospace vehicle utilizing the dipolar force field propulsion system can be propelled in the atmosphere of earth or vacuum of space. The propellant gas can be cryogenically cooled and be used for cooling superconducting magnets and can be boiled off and used as a propellant.

Also, the dipolar force field propulsion system of this invention can be combined with other known propulsion systems, such as a plasma propulsion system using hot ionized gases. By controlling the spatial angle between the E field and B field, the thrust of the dipolar force field propulsion system can be controlled.

In FIG. **36**, the alternating current high voltage is applied to the propulsion unit through a coil excitation transformer **850**. The coil excitation transformer establishes the B field in the conductive strips (**810**) in order to energize the embodiment described in connection with FIG. **35** and **36**.

The block diagram of FIG. **37** shows the alternating current power source for applying the alternating current power to the propulsion system. A high frequency oscillator **860** drives an amplifier **864** which has as an additional input thereto an alternating current power supply **862**. The amplifier applies the high voltage alternating current signal through the coil excitation transformer coupler **850** to drive the propulsion unit with the inductance and capacitance thereof shown as **866** and **868**, respectively.

FIGS. **38** and **39** show another embodiment for practicing the invention in the form of an VTOL vehicle. FIGS. **38** and **39** show an embodiment designed for VTOL utilizing a radio frequency inductive magnetic field. The field generates an azimuthal electric field which, in turn, generates a polarization current body force which acts vertically. The capacitance element for tuning the coil is incorporated into the vehicle's structure itself covering the full diameter of the vehicle and stores electrical energy which is cyclically converted into magnetic field energy of the coil as described in connection with the simple LCR tank circuit illustrated in FIG. **7**. The coil is a spiral winding which is formed by elements **900** which is supported in a vertical extending position as illustrated in FIG. **38** by an insulating strip **902**. The oapacitive surface is formed by upper outer surface **904** and inner surface **906** which is separated by an insulator **908**. The coil defined by elements **900** can be in the form of a spiral winding consisting of a number of turns, such as eight, which is parallel connected at both ends so as to reduce the equivalent inductance of the coil defined thereby. The coil is terminated at one end thereof by electrically connecting the same to one of the electrodes defining the capacitor, such as for example, electrode **904**, and the other end of the coil is connected to the other capacitor electrode such as inner capacitor electrode **906**. The coil is excited by an excitor coil **910** which is located in the center and driven by a high frequency generator **912** which is powered by a gas turbine engine **914** through a coupling clutch **916**. The generator **912** can be a superconductive generator which can generator ten kilowatts per kilogram of generator mass. A superconductor generator which is capable of generating this level of power is presently offered for sale by General Electric.

The exhaust from the turbine is exhausted through ports **920** which are defined by a shroud cover **922**.

The generator **912** also supplies electrical power to flashlamps **930** which are located beneath the vehicle and surrounded by a reflective surface **932**. The flashlamps **930** generate vacuum ultraviolet radiation in a controllable manner to excite the gas in selected regions underneath the field coil defined by windings **900**.

FIG. **39** is a top view, partially in section, which illustrates the spiral coil windings **900**. The coil consists of a flat ribbon conductors, preferably constructed as light as possible and formed of material such as aluminum alloy. The coil is electrically isolated via standoffs **942** from the high voltage plates formed by surfaces **904** and **906** which define the plates of the capacitor. The capacitor defined by the upper plate **904** and lower plate **906** is preferrably regularly slotted with slots **940** to prevent the formation of any eddy current losses due to the alternating magnetic field. The capacitor defined by the upper plate **904** and lower plate **906** provides structural support for the windings of the coils **900** through the insulating standoffs **942**. Thus, large magnetic pressures can be developed between the upper and lower surfaces **904** and **906** defining the capacitor, the insulating standoffs struts **942** and the windings **900**.

As illustrated in FIG. **39**, the air flows over the outer rim as well as through the central core which is indicated by arrows **960**. The air flow aids in collectively cooling the coil windings **900**.

FIGS. **40** and **41** illustrate a method of directional thrust control based upon an adjustable reflector **960**. The principle is illustrated diagramatically in FIGS. **42***a*, **42***b* and **42***c*. As long as the excitation source **962** radiation (here assumed to be flashtube) is symetrically distributed below and around the vertical axis of the vehicle **970** as shown in FIG. **42***a*, the thrust is vertical through the center of gravity of the vehicle. However, in FIG. **42***b*, if the field of radiation is shifted to one side, an increase or asymetry of excited states on that side of the vehicle exists resulting in increased thrust, which tilts the vehicle producing a horizonal thrust component moving the vehicle to the right. Moreover, the reflector **963** can be rotated through 360° in a plane parallel to the vehicle structure. Horizonal thrust component can be directed accordingly, as shown in FIG. **42***c*, where the reflector **962** has been rotated through 180°.

The construction of the directional control reflector **962** and ultraviolet radiation source **960** are more clearly understood by referring to FIG. **46** and **47**. Two flashtubes rotatably mounted about an axis **1000** below a platform **1002** supported by bearings **1004**. A gear wheel **1006** fixed to the vehicle structure. The flashtubes **960**, surrounded by reflectors **962** are adjustably mounted for rotation via a linear gear rack actuator **1012** acting upper semi-gear wheel **1014**. The power to the flashtubes is supplied via a pair of commutator rings **1016**, and connecting arm **1018**. The reflector **962** is rotatably mounted to the axis of the flashtube via spoke structure **1020**. The component effect is that the radiation field from the curved reflector **962** can be varied through 90 degrees of rotation about a horizonal axis from a horizonal plane to a verticle plane, as well as through 360 degrees about a vertical axis.

The vehicle VTOL dipolar propulsion system shown in FIG. **43** consists of a number of magnets arranged with their axis radially directed, with each alternate magnet of opposite polarity. The top field above the centerline of the magnets is shunted into the vehicle structure, without however effecting the external field below used to accelerate the ambient gas. This top field can now be used to bend a relativistic beam of electrons to produce ultraviolet (1000 A°) synchrotron radiation in the direction target to the beam, and, via an appropriate window and optical reflectors, direct the UV radiation into the gas for excitation of said gas.

The method of generating an alternating magnetic field is shown using D.C. superconductive magnets **1020** or permanent magnets. Use of the D.C. superconductive magnets with rotating ferrite shunts eliminates the A.C. current losses in the superconductive magnet arising therefrom due to resistence thereof, if the superconductive magents were operated in an A.C. mode to generate the same alternating magnetic field. The magent coils are arranged in a circle with alternate magnets in reversed field direction. A slotted ferrite disc **1040** rotating at high speed shunts the field of all magnets in one direction as shown in FIG. **44** leaving the unshunted field of the others expelled into the surrounding dielectric gas. The device is more clearly illustrated in FIG. **45** which graphically illustrates the magnitude and direction of the external field as the ferrite rotor **1040** is rotated by motor **1042** through several different angular positions. In **46***a*, the outward (north) positive fields of magnets **1030** are shunted through the ferrite leaving the inward (south) negative field unshunted and exposed to interact with the dielectric gas. As the ferrite rotor moves 22.5° to the position shown in FIG. **42**B, a neutral position is reached where the external field is approximately zero, as averaged, over the 45° of rotation. When the rotor **1040** reaches position shown as **42***c*, the ferrite shunts the onward (south) negative field, leaving the outward positive field exposed to the gas. Thus, through 90° of rotation the field has gone through a complete cycle of outward and then inward reversed field. The frequency of the alternating field is given by

$$f_r = N_r\,(\text{R.P.S.}) \tag{56}$$

where N_r is the number of magnet pairs of opposite polarity, and (R.P.S.) the frequency of rotation in revolutions per second (R.P.S.). The speed of rotation has been found by Beams to be limited to the rim velocity reaching the speed of sound of the material; where the centrifugal forces induce stresses sufficient to tear the rotor apart. Preferably, the ferrite rotor is reenforced with high strength material such as glass filaments. For example, a 1 meter diameter ferrite rotor spinning at 500 R.P.S. with 100 coil pairs mounted on a nonrotating frame could generate an intense field alternating at 50 kilocycles. For a fully excited gas (air at sea level), the thrust generated is sufficient to lift the vehicle even using rare earth magnets. The 500 R.P.S. or 30,000 R.P.M. could be generated by a gas turbine engine. Positive torque is required to break the magnetic field, but negative torque is obtained as the ferrite is attracted to the next coil. Hence, the average torque due to magnetic attraction is zero. The power is absorbed to reverse the field through the ferrite which has small losses since it is an insulator. Some heating is expected so air circulation is desirable to keep the rotors cool and prevent the superconductive magnets from heating up and going resistive.

FIGS. **47** and **48** illustrate a VTOL version of this method of field generation. The ferrite rotor rotates in a horizontal plane beneath the magnet coils arranged in a circle near the rim. A top row of ferrite plates fixed over the coils is used to shunt the field over the top of the vehicle which could produce an adverse downward force. The air gap between these plates and the coils is adjusted for this purpose. The radiation field used to excite the gas is derived from a free electron laser (FEL) **1050** using the same coils as the propulsion magnets **1052**. Electron guns are arranged near the rim of the top edge of the superconductive magnets **1052** and direct their electron beams **1054** in a circular path. The fields bend and accelerate the beam, **1054**. The acceleration produces synchrotron radiation in the far ultraviolet region which is directed to reflector **1060** which reflect the radiation **1062** downwards beneath the vehicle to excite the air. The excited air then interacts with the azimuthal electric fields produced by the alternated fields, is repelled downward, setting up a flow pattern around the vehicle as shown which generates the vertical thrust.

FIG. **48** illustrates pictorially the physical arrangement between the magnets **1052**, the shading magnets **1070** and the radiation **1062** traversing the magnets **1052** onto the reflector **1060**.

FIG. **49** is a pictorial representation of an aerospace vehicle using the dipolar force field propulsion system in combination with a rotating shunt plate **1092** and superconducting magnets **1090**. An appropriate energy source **100** is used for the aerospace vehicle. The radiation for exciting the particles **1104** is directed by reflectors **1102** to excite the gaseous atoms in the atmosphere under the spaceship. The electrical energy developed by the generator **1100** is rotatably coupled to the magnets through an electromagnetic coupling means **1108**.

HIGH ALTITUDE OPERATION

At high altitudes, the artificial excitation source can be deactivated and the natural ultraviolet radiation from the sun used to excite the air. Such phenomena is known in geophysics as "airglow," dayglow, nightglow and "aura borealis."

In addition to carrying power on board the vehicle for the purpose of exciting the gas around the vehicle, the gas may, to some extent, be excited from external sources such as a ground station or geosychronous power satellite. This has the distinct advantage of reducing weight. However, the frequencies are restricted to those which will propagate through the atmosphere with little attentuation, such as the visible and down to the microwave region; ultraviolet being highly absorbed. Thus the vehicle carries its own ultraviolet radiation source, such as from a syncrotron radiation source which can be varied to provide any desired distribution of wavelengths, e.g. by changing the energy of an electron beam. The absorbing frequency of the excited gas is given by the following equation for simple hydrogenic atoms:

$$\nu = CR_R\left(\frac{1}{n_l^2} - \frac{1}{n_u^2}\right) \quad (57)$$

where (nl) is the P.Q.N. of the lower state of interest and (n_u) is the upper state of interest. For excited states with n=40, and higher, the gas will absorb microwaves and increase the polarization, especially at higher altitudes where gas temperature and pressure is reduced. Thus, a ground station microwave source could enhance the polarization around a high flying electromagnetic aerospace vehicle.

In addition to absorption by electronic states, which enhances polarization for thrust augmentation purposes, other vibrational or rotational states may be created to absorb wavelengths of a specific nature to avoid reflection and consequent detection. This could be done automatically, by sensing the offending frequency, and adjusting the energy of the electron beam to control the spectral distribution of the synchroton radiation so as to excite the gases around the vehicle and absorb completely the offending frequency. If the frequency changes, the electron beam is likewise changed to again permit absorption of the offending frequency.

It is envisioned that the spacecraft illustrated in FIG. **49** could be operated in a vacuum, such as in interstellar space. It has been found by recent experiments that a momentum reaction force can be generated by the field itself due to the EXB vectors. This phenomenon is described in an article by G.M. Graham and D.G. Lahoz entitled "Observation of Static Electromagnetic Angular Momentum in Vacuum," *Nature*, Volume 285, May 15, 1980.

In FIG. **50**, a method of cyclically pumping an LCR tank circuit by magnetohydrodynamics so as to sustain the oscillations against the transfer of energy into the primary propulsion tank circuit is shown. The device consists of a rocket engine **(1160)** injected with fuel, oxygen and seed material to produce an electrically conducting plasma which passes through channel at velocity Vg with electrodes **1162** and field coils **1164** and ferrite core **1166** to increase magnetic permeability in the channel. The coils **1164** generate a varying current in series with the coils perpendicular to the plane of the paper, according to the equation:

$$\left. i = \frac{E}{R} = \frac{V_g}{R} B_0 \sin wt \right\} R = \text{resistance} \quad (58)$$

The current charges up the capacitor element C_s which discharges its current back into the coils at the resonant frequency that matches the primary circuit to the left.

The teachings of the invention have wide application. In its most generic application, the teachings can be utilized as a means for controllably accelerating a particle of matter having a selected dipole characteristic. Also, the invention teaches a method for controllably accelerating such a particle of matter.

The dipole force field propulsion system has utility for propelling an aerospace vehicle in the earth's atmosphere or in interstellar space. The propellant in the form of a cryogenic gas can be carried aboard the aerospace vehicle or the propellants can be external to but contiguous to the aerospace craft such as air or particles of matter or plasma in interstellar space. The energy sources likewise can be carried aboard the aerospace vehicle or can be external such as solar, microwave or laser excitation source.

What is claimed is:

1. A dipolar force field propulsion system comprising

means for generating an alternating electric field having its electromotive lines of force extending in a first direction and which vary at a selected frequency, said electric field having an electric field strength of a predetermined magnitude;

means for generating an alternating magnetic field having its magnetic lines of force which extend in a second direction which is at a predetermined angle to said first direction and which crosses and intercepts said electromotive lines of force at a predetermined location to define a spatial force field region and wherein the frequency of oscillation of the alternating magnetic field is substantially equal to the said selected frequency and is at a selected phase angle relating to said alternating electric field, said magnetic field having a flux density which when multiplied times the selected frequency is less than a known characteristic field ionization potential limit;

a source of neutral particles of matter having a selected electric dipole characteristic and having a known characteristic field ionization potential limit which is greater than said magnitude of the electric field strength, said dipoles of said particles of matter being capable of being driven into cyclic motion at said selected frequency by said electric field to produce a reactive thrust;

means for vaporizing said particles of the matter into a gaseous state at a selected temperature below the thermal ionization level thereof and for transporting said vaporized matter in said gaseous state into said spatial force field region defined by said crossed electromotive lines of force and magnetic lines of force which coact with and drive said dipoles into cyclic motion at said selected frequency to produce the reactive thrust which is substantially normal to said first direction of said electromotive lines of force and to the second direction of said magnetic lines of force; and control means operatively coupled to said means for generating an alternating electric field and to said means for generating an alternating magnetic field and which is responsive to the dielectric properties of the vaporized matter located in the spatial force field region for establishing a predetermined spatial and time relationship between the electric field, magnetic field and dipole cyclic motion for a selected frequency.

2. Means for generating a reactive thrust force adapted to propel an aerospace vehicle comprising

means for generating an alternating electric field which varies at a selected frequency and extends in a first direction, said electrical field having an electric field strength of a predetermined magnitude;

means for generating an alternating magnetic field at substantially said selected frequency which extends in a second direction which is positioned at a predetermined angle to said first direction and which crosses and intercepts said electric field at a predetermined location to define a force field region, said magnetic field having a flux density which when multiplied times said selected frequency is less than selected characteristic field ionization potential limit;

means for vaporizing neutral particles of matter into a gaseous state at a selected temperature which is below the thermal ionization level of said particle, said particle having a selected electrical dipole characteristic, a breakdown characteristic which is greater than the magnitude of the electric field strength and a selected characteristic field ionization potential limit, said dipoles of said matter being capable of being driven into cyclic motion at a selected frequency by said electric field;

means operatively coupled to said vaporizing means for transporting said particles of the material into said force field region wherein said crossing electric field and magnetic field coact with and cause said dipoles of matter to be driven into cyclic motion at substantially said selected frequency produce a reactive thrust force in a direction which is substantially perpendicular to said first direction and said second direction; and

control means operatively coupled to said electric field generating means and said magnetic field generating means and responsive to the dielectric properties of the vaporized particles of matter located in said force field region for establishing a predetermined spatial and time relationship between the alternating electric field, the magnetic field and frequency of the cyclic motion of said dipoles said control means establishing said selected frequency at substantially the resonant frequency of a capacitance and inductance circuit formed by the electric field generating means, said magnetic field generating means and the vaporized particles of matter in the force field region.

3. The propulsion system of claim **1** wherein the predetermined angle between the first direction of the electric field and the second direction of the magnetic field is selected to be 90°.

4. The propulsion system of claim **1** further comprising

means for raising the electronic excitation level of said particles of neutral matter in the vaporized gaseous state to a higher quantum level thereby increasing the magnitude of the selected electric dipole moment characteristic.

5. The propulsion system of claim **4** wherein said electronic excitation level raising means is a laser.

6. The propulsion system of claim **5** wherein said laser raises the quantum level of the particles of material to a quantum level between n=1 and n=20 by controlling the wavelength of the laser in step wise fashion to establish the dipole moment at a selected energy level which varies between the lowest energy level of the material at a quantum level of n=1 and a higher energy level which is below the thermal ionization level of the material at a quantum level of n=20.

7. The propulsion system of claim **4** wherein said electronic excitation level raising means is ultraviolet radiation.

8. The propulsion system of claim **5** wherein said laser originates from an external location.

9. The propulsion system of claim **1** wherein said magnetic field generating means are permanent magnets.

10. The propulsion system of claim **1** wherein said magnetic field generating means is a plurality of spaced, radially aligned superconducting magnets with the poles thereof alternately positioned relative to each adjacent magnet and a rotatable, ferrite magnetic material rotor which cyclically is transported past each of the radially aligned superconducting magnets to generate an alternating magnetic field.

11. The propulsion system of claim **1** wherein said magnetic field generating means is a coil.

12. A propulsion system comprising

a plurality of U-shaped superconducting pole pieces positioned in spaced alignment to define a substantially rectangular elongated channel;

a plurality of coils positioned one each around the center of one pole piece;

a plurality of pairs of spaced, substantially planar conducting electrodes positioned with each spaced pair extending between the ends of the U-shaped pole pieces wherein one of the planar electrodes is located adjacent the coil and the other planar electrode is spaced therefrom a distance substantially equal to the length of the U-shaped pole pieces defining a spatial region defined on two boundaries by the pair of planar electrodes and on two boundaries by the U-shaped end of the pole pieces, each of the spatial regions of each pole piece and planar electrode pair being in alignment enclosing said substantially elongated channel;

means for electrically connecting one of the planar electrodes of the planar electrode pair in series with the coil associated with its respective pole piece and for electrically connecting each series connected planar electrode pair and coil in parallel with the other series connected planar electrode pairs and being adapted to be connected to an alternating electrical power source;

a radiation source positioned at one end of the substantially rectangular elongated channel for directing radiation through each spatial region defined by each planar electrode pair and pole piece ends;

a plenum positioned at said one end adjacent said radiation source adapted to transport a vaporized propellent gas having neutral particles of matter having a selected dipole characteristic at a controlled rate through said substantially rectangular elongated channel; and a cryogenic source of propellant gas comprising neutral particles of matter wherein the particles of matter have a selected electric dipole characteristic and a known characteristic field ionization potential limit, said cryogenic source being operatively coupled to said plenum through a means for vaporizing said propellant gas to a level less than the ionization level thereof and applying a continuous stream of vaporized propellant gas to said plenum, said electrical connecting means being responsive to a said alternating electrical power source to produce an alternating electric field across each planar electrode pair and an alternating magnetic field between each pole piece and which was the alternating electric field establishing a plurality of aligned spatial force field region into which the vaporized gas is transported by said plenum into said substantially rectangular elongated channel through each spatial force field region of each planar electrode pair and coil, whereupon the particles of matter of the propellant gas are raised to an electronic excitation level by said radiation source and the crossing electric field and magnetic field which cause the dipoles of the particles of matter to be driven into cyclic motion to produce a reactive thrust.

13. The propulsion system of claim **12** wherein said radiation source is a laser.

14. The propulsion system of claim **12** wherein said means for vaporizing the propellant gas includes cooling means located adjacent each coil and pole piece center which is adapted to absorb heat from said coils which vaporizes said propellant gas passing therethrough.

15. The propulsion system of claim **14** further comprising

a flow meter positioned between said plenum and said vaporizing means to control the flow rate of the vaporized propellant gas into the plenum.

16. The propulsion system of claim **3** wherein said means for generating an alternating electric field is formed of a pair of spaced parallel plates which define a capacitor having a space between the parallel capacitive plates and wherein said means for generating an alternating magnetic field includes a plurality of spaced coils which are spaced relative to each other, said coils and said capacitor being electrically connected in series, resulting in the electric field being phase displaced from the magnetic field by 90°, said magnetic means being adapted to direct and concentrate the lines of magnetic force between the capacitive plates to establish an electric field which is located at substantially **90°** relative to the magnetic field.

17. The propulsion system of claim **16** wherein said means for generating the alternating magnetic field includes a ferrite coil for further concentrating said magnetic lines of flux.

18. The means for generating a reactive thrust force of claim **2** wherein said means for generating an alternating current electric field is a pair of spaced parallel electrodes which are adapted to distribute charges on a controlled surface and to distribute the charges uniformly on the surface thereof to produce an electric field which extends in said first direction and which is adapted to intercept the magnetic field generated by said means for generating throughout the spatial region in the vicinity of the electrode an alternating magnetic field at substantially right angles to form a crossed dipole magnetic throughout said spatial region field which is substantially at a right angle to the electric dipole field.

19. The dipolar force field propulsion system of claim **1** further comprising

means including an excitation source of radiation positioned at a selected location on the system for producing a field of radiation; and

means including means defining a reflecting surface postion adjacent radiation field producing means for selectively positioning said reflective surface at a controlled angle relative to said excitation source to produce a thrust component of force in a selected direction causing said system to move in a direction opposite to said selected direction of a thrust component of force.

20. The propulsion system of claim **4** further comprising

a source of second neutral particles of matter which is of different species than the source of neutral particles of matter and wherein the second neutral particles of matter have a selected electron dipole characteristics, said second neutral particles of matter being capable of interacting with said neutral particles of matter which have been raised to an electronic excitation level enabling the atoms of said second neutral particles of matter to act as buffer atoms with said neutral particles of matter to permit optical pumping of the neutral particles of matter at raised electronic excitation levels.

21. The propulsion system of claim **20** further comprising

a laser adapted to optically pump said neutral particles of matter.

* * * * *

INITIAL POWER SYSTEMS

I HAVE INCLUDED THE FOLLOWING TWO PATENTS TO ADDRESS THE ISSUE OF INTERNAL POWER.

THE FIRST PATENT COMES FROM VOLUME V OF THE UFO HOW-TO SERIES "FUSION AND ANTI-MATTER" AND WOULD MAKE A GOOD SOURCE OF ENERGY AND PROPULSION.

THE SECOND PATENT IS INCLUDED AS EITHER A FAILSAFE MECHANISM, OR AS A STARTER SYSTEM, TO WHATEVER THE BUILDER OF THE CRAFT WOULD PREFER. IT IS AN EXPIRED PATENT FROM BRITAIN. THE U.S. WOULD NEVER GRANT THIS PATENT SIMPLY BECAUSE OF THE TITLE: "METHOD AND MEANS OF PRODUCING PERPETUAL MOTION WITH HIGH POWER."

REGARDLESS OF THE TITLE, AN EFFECTIVE SYSTEM IS AN EFFECTIVE SYSTEM; THERE IS NO ARGUING WITH SUCCESS.

YOU CAN LEARN MORE ABOUT COLD FUSION AND THE RELATED SCIENCE IN VOLUME V OF THE UFO HOW-TO SERIES: "FUSION AND ANTI-MATTER."

PCT WORLD INTELLECTUAL PROPERTY ORGANIZATION

INTERNATIONAL APPLICATION PUBLISHED UNDER THE PATENT COOPERATION TREATY (PCT)

International Patent Classification 5 : G21B	A2	International Publication Number: WO 90/14668 International Publication Date: 29 November 1990 (29.11.90)

International Application Number: PCT/US90/02424

International Filing Date: 1 May 1990 (01.05.90)

Priority data:
347,473 4 May 1989 (04.05.89) US

Applicant and Inventor: CRAVENS, Dennis, J. [US/US]; 2222 Wheeler Street, Vernon, TX 76384 (US).

Agent: ETHINGTON, Paul, J.; Reising, Ethington, Barnard, Perry & Milton, Post Office Box 4390, Troy, MI 48099 (US).

Designated States: AT (European patent), BE (European patent), CA, CH (European patent), DE (European patent)*, DK (European patent), ES (European patent), FR (European patent), GB (European patent), IT (European patent), JP, LU (European patent), NL (European patent), SE (European patent), SU.

Published
Without international search report and to be republished upon receipt of that report.

Title: COLD FUSION PROPULSION APPARATUS AND ENERGY GENERATING APPARATUS

Abstract

Propulsion apparatus employs "cold fusion" of deuterium absorbed in a metal host lattice (7) to generate a heated momentum exchange effluent stream from the deuterium itself and/or to heat a momentum exchange fluid to provide a propulsive impulse upon exhausting through a nozzle (9). Thermal efficiency of a propulsion apparatus and an energy generating apparatus employing such "cold fusion" of deuterium in a metal host lattice (7) is improved by using a deuterium (hydrogen)-absorbing metal lattice alloyed or compound with one or more of W, Re, Mo, Tu, Ti, Ir and C to raise the lattice melting point and permit higher "cold fusion" temperatures.

DESIGNATIONS OF "DE"

Until further notice, any designation of "DE" in any international application whose international filing date is prior to October 3, 1990, shall have effect in the territory of the Federal Republic of Germany with the exception of the territory of the former German Democratic Republic.

FOR THE PURPOSES OF INFORMATION ONLY

Codes used to identify States party to the PCT on the front pages of pamphlets publishing international applications under the PCT.

AT	Austria	**ES**	Spain	**MC**	Monaco
AU	Australia	**FI**	Finland	**MG**	Madagascar
BB	Barbados	**FR**	France	**ML**	Mali
BE	Belgium	**GA**	Gabon	**MR**	Mauritania
BF	Burkina Fasso	**GB**	United Kingdom	**MW**	Malawi
BG	Bulgaria	**GR**	Greece	**NL**	Netherlands
BJ	Benin	**HU**	Hungary	**NO**	Norway
BR	Brazil	**IT**	Italy	**RO**	Romania
CA	Canada	**JP**	Japan	**SD**	Sudan
CF	Central African Republic	**KP**	Democratic People's Republic of Korea	**SE**	Sweden
CG	Congo			**SN**	Senegal
CH	Switzerland	**KR**	Republic of Korea	**SU**	Soviet Union
CM	Cameroon	**LI**	Liechtenstein	**TD**	Chad
DE	Germany, Federal Republic of	**LK**	Sri Lanka	**TG**	Togo
DK	Denmark	**LU**	Luxembourg	**US**	United States of America

COLD FUSION PROPULSION APPARATUS AND ENERGY GENERATING APPARATUS

Field Of The Invention

The present invention relates to power generation and propulsion and, in particular, to "cold fusion" power generating devices and propulsion devices.

Background Of The Invention

Specific impulse is the measure of thrust generated per weight (mass) of propellant used. Current chemical propulsion devices (rockets) are approaching their theoretical limits in terms of specific impulse. For example, one of the more successful chemical propulsion systems includes liquid oxygen and hydrogen and exhibits a specific impulse of generally 400 to 500 sec. One of the limitations of such a chemical system is that, for a given amount of energy or heat available for the chemical reaction, the specific impulse is limited, in part, by the molecular weights of the effluent gases.

Another major constraint on chemical propulsion systems is that systems which have a high specific impulse inherently have a low thrust and, likewise, systems with high thrust normally exhibit a low specific impulse. This situation is unfortunate since high thrust is required in the lower stages of a space mission but high specific impulse is beneficial in subsequent higher stages (in space or orbit) of the mission. For example, high thrust is required to lift the system from the earth while, once lifted or in orbit, only low thrust is required and is advantageous from a fuel efficiency standpoint. Unfortunately, provision of high specific impulse in the upper stages of a mission lowers the available thrust in the lower stages of the mission.

What is needed is a propulsion apparatus which does not suffer from these disadvantages. In particular, a propulsion apparatus not solely of the chemical type and thus not limited by the molecular weight of the effluent products would be desirable. Also, a propulsion system which exhibits a higher temperature of effluent products to increase specific impulse would be desirable. Moreover, a propulsion system which exhibits both high thrust and high specific impulse or, alternatively, a variable thrust and specific impulse, would be desirable.

With respect to power generating apparatus, recent work by Stanley Pons, University of Utah, and Martin Fleischmann, the University South Hampton, indicates that a metal host lattice of palladium can absorb deuterium ions in an electrolytic cell where the palladium is rendered cathodic in an aqueous solution of deuterium oxide and LiOH. The confinement of the deuterium boseons within the Fermi sea of metal electrons appears to promote a nuclear transformation/reaction (currently known as "cold fusion") where tritium and helium isotopes are released along with heat and other energetic forms, although the exact physical theory is uncertain at present. To date, the particular type of reaction product/energy emissions from the palladium cathode have varied depending upon the experimental configuration employed.

For some applications, especially for applications in space environments, the thermal efficiency of such a "cold fusion" power generating apparatus is extremely important since any inefficiencies appear as rejected heat which is extraordinarily difficult to discard in space environments. Since the thermal efficiency of a power generating apparatus is governed by the temperature differentials due to thermodynamic constraints, it would be desirable to provide a "cold fusion" power generating apparatus with a relatively high operating temperature and thus a high thermal efficiency for particular use in space environments. Moreover, it would be desirable to supply deuterium to the metal host lattice in a more practical manner not limited by the boiling point of "heavy" water (e.g., 101°C) used heretofore by Pons and Fleischmann in practicing the "cold fusion" process.

It is an object of the present invention to satisfy these needs and desires.

SUMMARY Of The Invention

The invention contemplates a propulsion apparatus employing "cold fusion" of deuterium absorbed in a metal host lattice to generate a heated momentum exchange effluent stream directly from the deuterium itself and/or to heat a momentum exchange fluid flowing relative to the metal host lattice to provide a propulsive impulse when exhausted.

In one embodiment of the invention, a propulsion apparatus employs "cold fusion" of deuterium absorbed in a metal host lattice to generate a momentum exchange effluent stream therefrom comprising any heated unconsumed (unfused) deuterium and products resulting from nuclear transformation of the absorbed deuterium and means for exhausting the effluent stream to provide a propulsive impulse.

In another embodiment of the invention, a propulsion apparatus employs "cold fusion" of deuterium absorbed in a metal host lattice to generate heat in the metal lattice for heating a momentum exchange fluid (e.g., a gas, liquid, plasma, etc.) flowing through one or more passages in the metal lattice and means for exhausting the heated fluid to provide a propulsive impulse. The momentum exchange fluid preferably has a molecular weight below 50; e.g., hydrogen, oxygen, air, water, helium, ammonia or carbon dioxide and mixtures thereof.

In another embodiment of the invention, first and second reactive fluids (e.g., preferably hydrogen and oxygen) are supplied to respective first and second passage means through one or more metal host lattices capable of absorbing deuterium for the generation of heat in the metal lattice by nuclear transformation of the deuterium therein. The first and second fluids are heated as they pass through the respective first and second passage means. Means is provided for mixing the heated first and second fluids such that they react exothermally. The reaction products are then exhausted through suitable means, e.g., through a nozzle, to provide a propulsive impulse.

In the aforesaid embodiments of the inventive propulsion apparatus, means is preferably provided between the metal host lattice and the source of deuterium and/or the source of momentum exchange fluid for regulating the quantity (or flow) of deuterium and/or momentum exchange fluid for the purpose of varying the thrust and specific impulse of the propulsion apparatus.

The invention also envisions improving the thermal efficiency of a propulsion apparatus and an energy producing apparatus employing "cold fusion" of deuterium absorbed in a metal host lattice by virtue of supplying the deuterium as deuterium gas or plasma to a deuterium (hydrogen)-absorbing metal lattice alloyed or compounded with one or more of W, Re, Mo, Ta, Ti, Ir and C to raise the melting point of the metal lattice to permit higher "cold fusion" temperatures in the metal lattice.

In the aforesaid bodiments of the invention, the metal host lattice can be rendered cathodic relative to a source of deuterium gas or plasma or a deuterium-bearing fused salt electrolyte. The electrical potential applied to the metal host lattice may be varied to control and throttle the "cold fusion" of deuterium in the metal host lattice.

Brief Description Of The Drawings

Figures 1-5 are schematic illustrations of various embodiments of the invention.

Detailed Description Of The Invention

Fig. 1 illustrates one embodiment of the invention using deuterium directly as a momentum exchange effluent stream or medium. In particular, a liquefied deuterium storage vessel 1 is connected to a deuterium gas reservoir 3 through a control valve 2 to supply deuterium to the reservoir 3. Pressurized deuterium gas (e.g., at 1000 psi) is released from the reservoir 3 through a regulating valve 6 for supply to the upstream side 7a of a deuterium (hydrogen)-absorbing metal host lattice 7 described hereinbelow for absorption into the metal lattice 7. An optional electrical connection can be made from either the reservoir 3 or a grid 4 within the reservoir 3 to a positive terminal V+ of a DC voltage supply 5 while the negative terminal V- of the D.C. voltage supply 5 is connected to the metal host lattice 7. However, with the proper selection of deuterium gas temperatures, flow rates, pressures and the metal of metal host lattice 7, deuterium absorption into the metal lattice 7 is achievable without use of the DC voltage supply 5 (i.e., it may be eliminated) so as to avoid space charge repulsion in the deuterium effluent stream in the exhaust nozzle 9.

As mentioned, pressurized deuterium gas (e.g., at 1000 psi) is released from the reservoir 3 through the regulating valve 6 for supply to the upstream side 7a of the metal host lattice 7 for absorption therein. Other isotopes of hydrogen can be used in conjunction with the deuterium. The deuterium gas may be supplied by a pulsing technique wherein the valve 6 is opened until a deuterium gas pressure build-up is detected (as a result of gas heat-up by cold fusion in the metal host lattice) and then shut off until the pressure drops to selected level whereupon the valve is again opened. A nuclear transformation/reaction of the absorbed deuterium occurs in the metal host lattice 7 in accordance with the aforementioned "cold fusion" process to generate energy (and heat) in the metal host lattice 7. Any heated unconsumed deuterium and products of the nuclear transformation/reaction of deuterium (e.g., tritium and helium isotopes) are allowed to exhaust (i.e., to escape and expand) from the downstream side 7b of the metal host lattice 7 as a momentum exchange effluent stream or medium and are conducted through the exhaust nozzle 9 (which is designed to match the Mach considerations as known in the art) to provide a propulsive impulse or force. The nozzle 9 can be used to vector the effluent stream.

Use of any unconsumed deuterium and the nuclear transformation products of deuterium directly as the effluent stream in the above-described propulsive apparatus of the invention provides a simple and yet effective means for providing a propulsive impulse. The metal host lattice 7 is made so that diffusion of the absorbed deuterium through the entire thickness thereof can be achieved within a reasonable time frame. For example, the time must be long enough for confinement-induced fusion ("cold fusion") to be initiated and maintained in the metal host lattice 7 yet short enough for any unconsumed (unfused) absorbed deuterium and the aforementioned nuclear transformation products to escape or expel from the downstream side 7b of the lattice 7 as an effluent stream.

Use of deuterium gas (or plasma as will be described herebelow) avoids a severe temperature limitation placed on the "cold fusion" process when an aqueous solution of deuterium oxide and LiOH is used; i.e., the temperature of the process is not limited by the boiling point of heavy water (101°C) at ambient pressure. As a result, the invention envisions using the "cold fusion" process in the propulsive apparatus described hereinabove at much higher temperatures, such as preferably between about 1000°C and about 3000°C. An embodiment of the invention where the metal host lattice 7 achieves a temperature of 1000°C is expected to provide a specific impulse of near 800 sec, which constitutes a gain of 2 or more over the specific impulse achievable with conventional chemical systems.

In the embodiment of the invention described hereinabove as well as hereinbelow, the metal host lattice 7 may comprise a Pd sheet approximately 2mm to 6mm in thickness. The Pd sheet can be formed to present a large surface area on upstream side 7a to the deuterium gas released from the reservoir 3 and on the downstream side 7b to the exhaust nozzle 9 to increase through-put and thus the thrust level of the propulsion apparatus. Other configurations of the metal host lattice 7 can be used in practicing the invention.

In lieu of palladium as the metal host lattice 7, other deuterium (hydrogen)-absorbing metals can be used, such as, for example, Ti, Ni and Mg and alloys thereof one with another (e.g., Ni with 10 w/o Mg). Preferably, however, in accordance with another aspect of the invention, the metal host lattice 7 comprises a deuterium (hydrogen)-absorbing metal, preferably one or more of the metals Ti, Pd, Ni and Mg, alloyed or compounded with one or more of W, Re, Mo, Ta, Ti, Ir and C to provide a metal host lattice having a melting temperature of about 1800°C and above. When carbon is used to raise the melting temperature, the metal host lattice 7 may comprise a carbide of one or more of the aforementioned metals. When the metal host lattice 7 is made of these higher melting point alloys/compounds, correspondingly higher "cold fusion" temperatures can be achieved (e.g., about 1000°C to about 3000°C) and will result in enhanced thermal efficiency and the possibility of even greater specific impulses.

In lieu of using deuterium gas as a source for absorption into the metal host lattice 7, the invention contemplates using a deuterium plasma or deuterium-bearing fused salt electrolyte to provide a supply of deuterium to the metal host lattice 7. For example, in Fig. 2, a conventional plasma generator 30 is shown downstream of the deuterium gas reservoir 3 and the valve 6 to establish a deuterium plasma for contacting the upstream side 7a of the metal lattice 7 to introduce absorbed deuterium into the metal lattice 7. In Fig. 3, a fused salt electrolyte (e.g., LiH/LiD) 31 is shown as a source of deuterium adjacent the upstream side 7a of the metal lattice 7 for the same purpose. In these figures, like features of Fig. 1 are represented by like reference numerals.

Although the D.C. power supply 5 is optional in the aforementioned embodiments of the invention , it may be possible to employ the power supply 5 as a means of controlling/throttling the "cold fusion" reaction in the metal lattice 7; e.g., by varying the electrical potential applied on the metal lattice 7 to control the quantity of electrons/absorbed deuterium in the metal lattice.

A preferred propulsion apparatus for space applications is shown in Fig. 4 and exhibits both high thrust and high specific impulse or, alternatively, a variable thrust and variable specific impulse. In Fig. 4, like features of Fig. 1 are represented by like reference numerals primed.

In this embodiment of the invention, a low molecular working fluid is disposed in a storage vessel 11' and functions as a momentum exchange fluid (working fluid). The storage vessel 11' is connected through a valve 12′ to a passage 14′ so as to flow the working fluid through the passage 14′. As shown, the passage 14′ is formed within a suitable high temperature tubular member 13′, such as, for example, a W or Al_2O_3 tube, disposed in the metal host lattice 7′. The material of the tubular member 13′ is selected to have a reasonably high thermal conductivity and a low permeability to deuterium absorbed in the metal lattice 7′ to substantially prevent absorbed deuterium from mixing with the momentum exchange fluid passing through the passage 14′. The tubular member 13′ is in direct thermal contact with the metal host lattice 7′ so as to transfer heat to the momentum exchange fluid as it passes through the passage 14′. The tubular member 13′ and the passage 14′ may be of any suitable shape. The heated momentum exchange fluid is allowed to expand and exhaust through nozzle 9′ to provide a propulsive impulse.

Deuterium gas under pressure (e.g., 1000 psi) is supplied to the metal host lattice 7′ from storage vessel 1′ through a valve 2′ as described hereinabove for the embodiment of Fig. 1. Similarly, the metal lattice 7′ can be biased cathodically (negatively relative to the deuterium gas source) by

D.C. power supply 5′ as described hereinabove for the embodiment of Fig. 1. Heat is generated in the metal host lattice 7′ by "cold fusion" of the deuterium absorbed therein.

In the embodiment of Fig. 4, the metal lattice 7′ (i.e., the "cold fusion" energy generating means) is isolated from the momentum exchange fluid by the tubular member 13′. The tubular member 13′ (or other means for forming one or more passages) is so provided as to form one or more passages through the metal lattice 7′ for passage of the momentum exchange fluid and yet provide an interface through which the absorbed deuterium cannot pass into the momentum exchange fluid. In this way, the momentum exchange fluid can absorb thermal energy from the metal lattice 7′ without disrupting the "cold fusion" reaction/transformation in the metal lattice 7′. At the same time, the momentum exchange fluid can be throttled by valve 12′ to vary both the thrust and specific impulse of the propulsion apparatus.

Since the available thermal energy is limited by the available fusion energy, the temperature reached by the momentum exchange fluid will be limited by the flow rate of the momentum exchange fluid through the passage 14'. At higher flow rates, higher thrust will be achieved at the expense of lower specific impulse (i.e., lower average energy available to the exhausted fluid). At lower flow rates, higher specific impulse will be achieved with a corresponding decrease in thrust. Momentum exchange fluids having a low molecular weight, preferably below 50, are preferred. Preferred low molecular weight momentum exchange fluids for use in the embodiment of Fig. 4 include H_2, O_2, air, H_2O, He, NH_3, and CO_2 and mixtures thereof. The first six of these (i.e., H_2, O_2, air, H_2O, He, NH_3,) have desirable trade-offs between low molecular weight and storage properties. The latter of these (i.e., CO_2) has desirable storage properties (e.g., storable as dry ice) and is easily obtained both on earth and on Mars, thus leading to interplanetary uses by spacecraft.

Fig. 5 illustrates still another embodiment of the invention for providing a high thrust fusion/chemical propulsion apparatus. This embodiment includes first and second storage vessels 50,52 for first and second chemically reactive momentum exchange fluids (working fluids); e.g., preferably H_2 in vessel 50 and O_2 in vessel 52. The momentum exchange fluids are flowed through respective flow control valves 54,56 and through respective passages 60,62 formed by tubular members 70,72 in thermal contact with respective first and second metal host lattices 80,82; e.g., in the same fashion as described hereinabove for Fig. 4. Pressurized deuterium gas (e.g., 1000 psi) is supplied to each metal lattice 80,82 from a common storage vessel 90 via respective valves 100,102 such that each metal lattice 80,82 absorbs deuterium and generates thermal energy via the "cold fusion" process described for the embodiments of Fig. 1-4. The first and second reactive momentum exchange fluids are heated as they flow through the passages 60,62 and the fluids at elevated temperature are then conducted to a mixing chamber 110 where they can chemically react in exothermic fashion upon mixing. The reaction products are allowed to expand and escape from the mixing chamber 110 through the exhaust nozzle 112 to provide a propulsive impulse.

In this way, the temperature of the reaction products (exhausted through nozzle 112) can be increased to a level which is not achievable by either fusion or chemical reactions alone. In particular, the fusion energy supplied to momentum exchange fluids flowing through passages 60,62 is additive to the energy available from the exothermic chemical reaction to raise the temperature of the reaction products exhausted through the nozzle 112. To this end, the first and second momentum exchange fluids (in vessels 50,52) should be chosen so that the reaction products in the mixing chamber 110 do not disassociate or are not at lower energy levels. Although H_2 and O_2 are described hereinabove as the momentum exchange fluids, other momentum exchange fluids, which are stable at elevated temperatures and result in stable reaction products, may be used.

Moreover, the energy available from the fusion process can allow access to chemical reactions that would normally be unavailable. For example by elevating the energy states of either or both working fluids, the reaction times can be reduced to the benefit of the nozzle design. Also solid materials, such as Li, can be liquefied or vaporized to react with the second fluid, for example oxygen. In this latter case, solid Li greatly reduces the storage volume and the weight of the associated storage vessel.

While certain preferred embodiments of the invention have been described hereinabove, those skilled in the art will recognize that various modifications and changes can be made therein for practicing the invention as described in the following claims.

I Claim

1. A propulsion apparatus for rocket, spacecraft and terrestrial use, comprising

(a) means for providing deuterium to a metal host lattice capable of absorbing said deuterium for the generation of heat within the metal lattice by nuclear transformation of the deuterium therein, and

(b) means for exhausting as an effluent stream from the metal lattice any heated unconsumed deuterium and products of the nuclear transformation to provide a propulsive impulse.

2. The apparatus of claim 1 wherein said means for providing deuterium to the metal host lattice comprises means for delivering deuterium gas or plasma to the metal host lattice.

3. The apparatus of claim 2 wherein the metal host lattice is rendered cathodic relative to a source of the deuterium gas or plasma.

4. The apparatus of claim 1 wherein said means for providing deuterium to the metal host lattice comprises a fused salt electrolyte.

5. The apparatus of claim 1 wherein the metal host lattice comprises one or more of the metals Ti, Pd, Ni and Mg.

6. The apparatus of claim 1 wherein the metal host lattice comprises a deuterium-absorbing metal alloyed or compounded with one or more of W, Re, Mo, Ta, Ti, Ir and C to increase the melting point of said metal host lattice, thus allowing a higher temperature to be generated in said metal host lattice.

7. The apparatus of claim 1 further including means for regulating the quantity of deuterium provided to the metal host lattice for varying the thrust and specific impulse of the propulsion apparatus.

8. The apparatus of claim 1 wherein said means for exhausting includes a nozzle means through which any heated unconsumed deuterium and the reaction products exit to provide the propulsive impulse.

9. The apparatus of claim 2 further including a holding tank for holding pressurized deuterium gas for delivery to the metal host lattice as controlled by valve means between said holding tank and said metal host lattice.

10. A propulsion apparatus for rocket, spacecraft and terrestrial use, comprising:

(a) means for providing deuterium to a metal host lattice capable of absorbing said deuterium for generation of heat within the metal lattice, the metal lattice having passage means,

(b) means for supplying a momentum exchange fluid to the passage means to heat said momentum exchange fluid; and

(c) means for exhausting the heated momentum exchange fluid from the passage means to provide a propulsive impulse.

11. The apparatus of claim 10 wherein said passage means is formed by a deuterium-impermeable hollow member disposed in said metal host lattice for transferring heat from the metal host lattice to the momentum exchange fluid.

12. The apparatus of claim 10 wherein the momentum exchange fluid has a molecular weight less than 50.

13. The apparatus of claim 12 wherein the momentum exchange fluid comprises one or more of hydrogen, oxygen, air, water, helium, ammonia and carbon dioxide.

14. The apparatus of claim 10 wherein said means for providing deuterium to the metal host lattice comprises means for delivering deuterium gas or plasma to the metal host lattice.

15. The apparatus of claim 10 further including means for regulating the flow of the momentum exchange fluid to the passage means to vary the thrust and specific impulse of the propulsion apparatus.

16. The apparatus of claim 10 wherein the metal host lattice comprises one or more of the metals Ti, Pd, Ni and Mg.

17. The apparatus of claim 10 wherein the metal host lattice comprises a deuterium-absorbing metal alloyed or compounded with one or more of W, Re, Mo, Ta, Ti, Ir and C to increase the melting point of said metal host lattice, thus allowing a higher temperature to be generated in said metal host lattice.

18. The apparatus of claim 10 wherein said means for exhausting includes a nozzle means through which any heated unconsumed deuterium and nuclear reaction products thereof pass to provide the propulsive impulse.

19. The apparatus of claim 10 further including a holding tank for holding pressurized deuterium gas for delivery to the metal host lattice as controlled by valve means between said holding tank and said metal host lattice.

20. A propulsion apparatus, comprising:

(a) metal lattice means capable of absorbing deuterium for the generation of heat therein and having a first passage means and a second passage means,

(b) means for providing deuterium to said metal lattice means,

(c) means for supplying a first reactive fluid and second reactive fluid to the respective first passage means and the second passage means to heat the first fluid and the second fluid, the first fluid and the second fluid being reactive in exothermic manner at elevated temperature,

(d) means for mixing the first fluid and the second fluid after they are heated so as to exothermically react the first fluid and the second fluid, and

(e) means for exhausting the reaction products of the reacted first fluid and the second fluid to provide a propulsive impulse.

21. The apparatus of claim 20 wherein said metal lattice means comprises a first metal lattice and a second, separate metal lattice and wherein said first passage means is in said first metal lattice and said second passage is in said second metal lattice.

22. The apparatus of claim 20 wherein the first fluid is hydrogen and the second fluid is oxygen.

23. The apparatus of claim 20 wherein said mixing means comprises a mixing chamber connected to said first passage means and said second passage means for receiving the first fluid and second fluid after heating.

24. A thermal energy generating apparatus, comprising:

(a) a metal host lattice comprising a deuterium-absorbing metal alloyed or compounded with one or more of W, Re, Mo, Ta, Ti, Ir and C, and

(b) means for providing deuterium to the metal host lattice for absorption into the metal host lattice and generation of heat therein by nuclear transformation of the deuterium therein.

25. The apparatus of claim 24 wherein said means for providing deuterium comprises means for delivering deuterium gas or plasma to the metal host lattice.

26. The apparatus of claim 24 wherein said means for providing deuterium comprises a deuterium-bearing fused salt electrolyte.

27. A method of generating a propulsive impulse, comprising the steps of:

(a) supplying deuterium to a metal host lattice for absorption therein and generation of heat within the metal host lattice by nuclear transformation of said deuterium, and

(b) exhausting as an effluent stream from the metal lattice any heated unconsumed deuterium and products of the nuclear transformation so as to provide a propulsive impulse.

28. A method of generating a propulsive impulse, comprising the steps of:

(a) supplying deuterium to a metal host lattice for absorption therein and generation of heat within the metal host lattice, and

(b) passing a momentum exchange fluid through one or more passages in the metal host lattice to heat the momentum exchange fluid, and

(c) exhausting the heated momentum exchange fluid from the passage as to generate a propulsive impulse.

29. A method of generating a propulsive impulse, comprising the steps of:

(a) supplying deuterium to a metal host lattice means for absorption therein and generation of heat within the metal host lattice, and

(b) flowing a first reactive fluid and a second reactive fluid through respective first and second passages in the metal host lattice means to heat the first reactive fluid and the second reactive fluid,

(c) exothermically reacting the heated first reactive fluid and the heated second reactive fluid, and

(d) exhausting the reaction products of the reacted first reactive fluid and the second reactive fluid to generate a propulsive impulse.

30. A method of generating energy comprising, supplying deuterium as a gas or plasma to a metal host lattice comprising a deuterium-absorbing metal alloyed or compounded with one or more of W, Re, Mo, Ta, Ti, Ir and C and transforming the deuterium in the metal host lattice to generate heat.

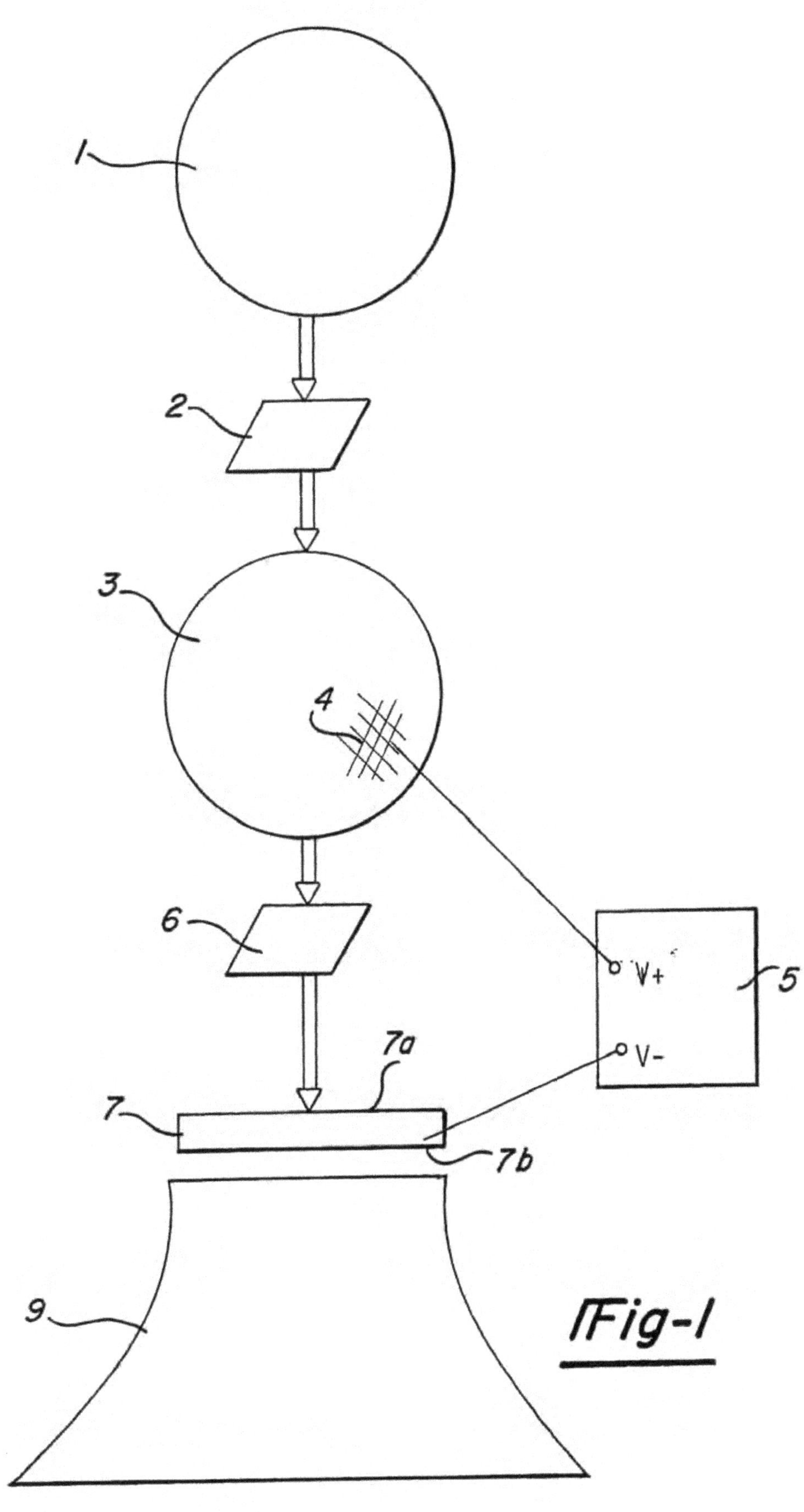

Fig-1

SUBSTITUTE SHEET

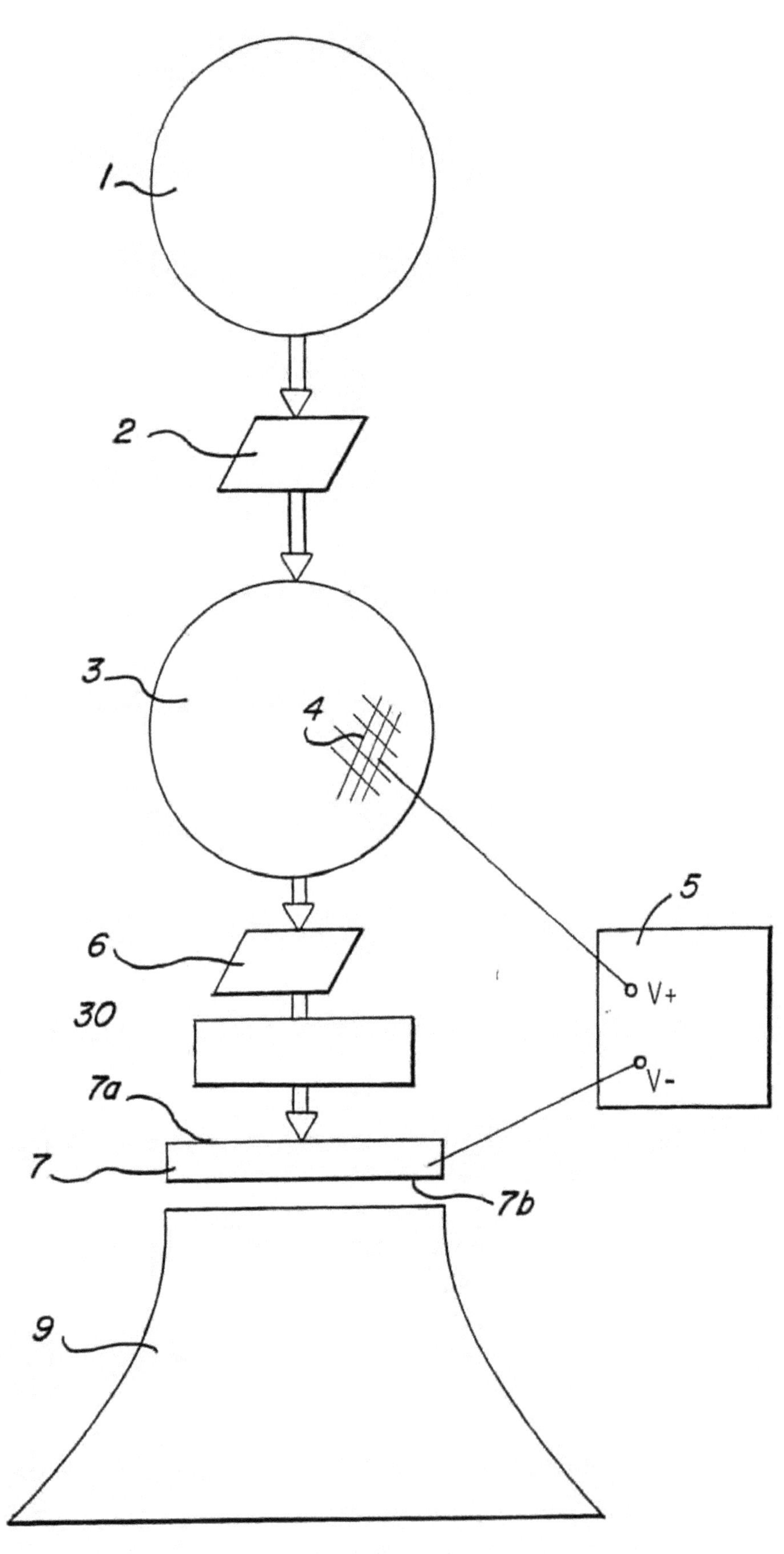

IFig-2

SUBSTITUTE SHEET

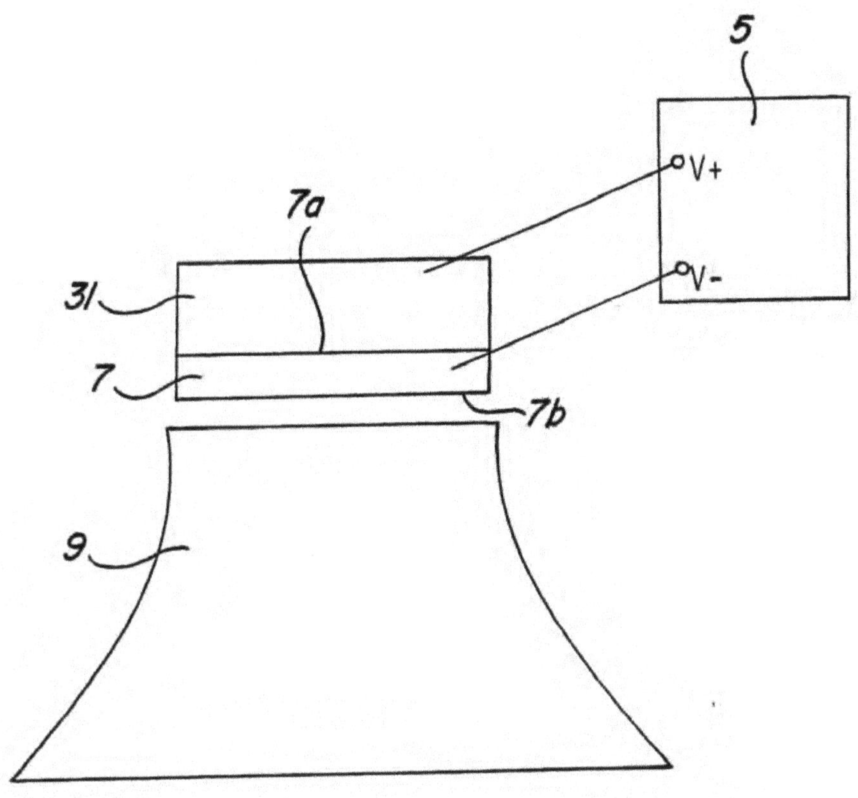

Fig-3

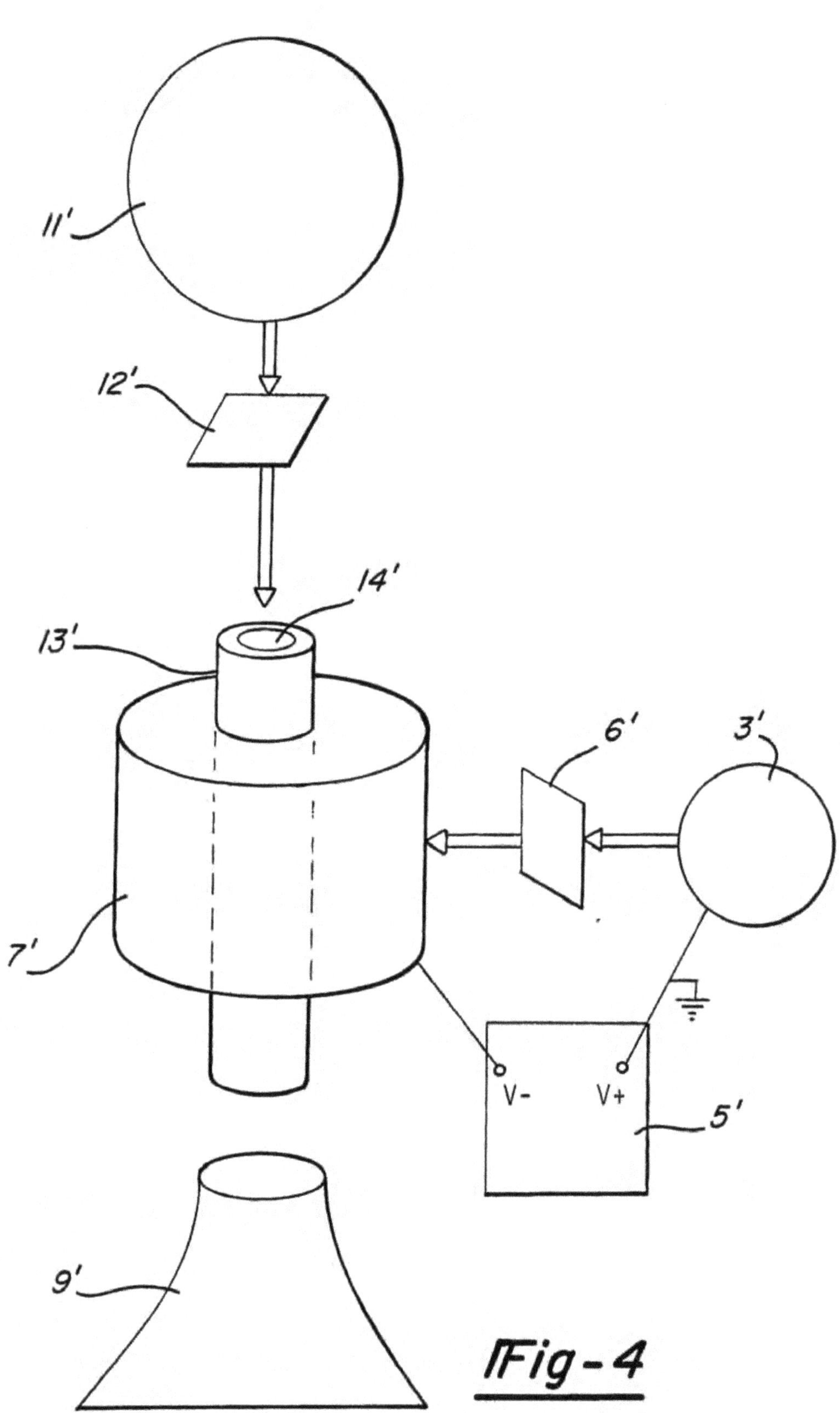
11'
12'
14'
13'
6'
3'
7'
V-
V+
5'
9'
Fig-4

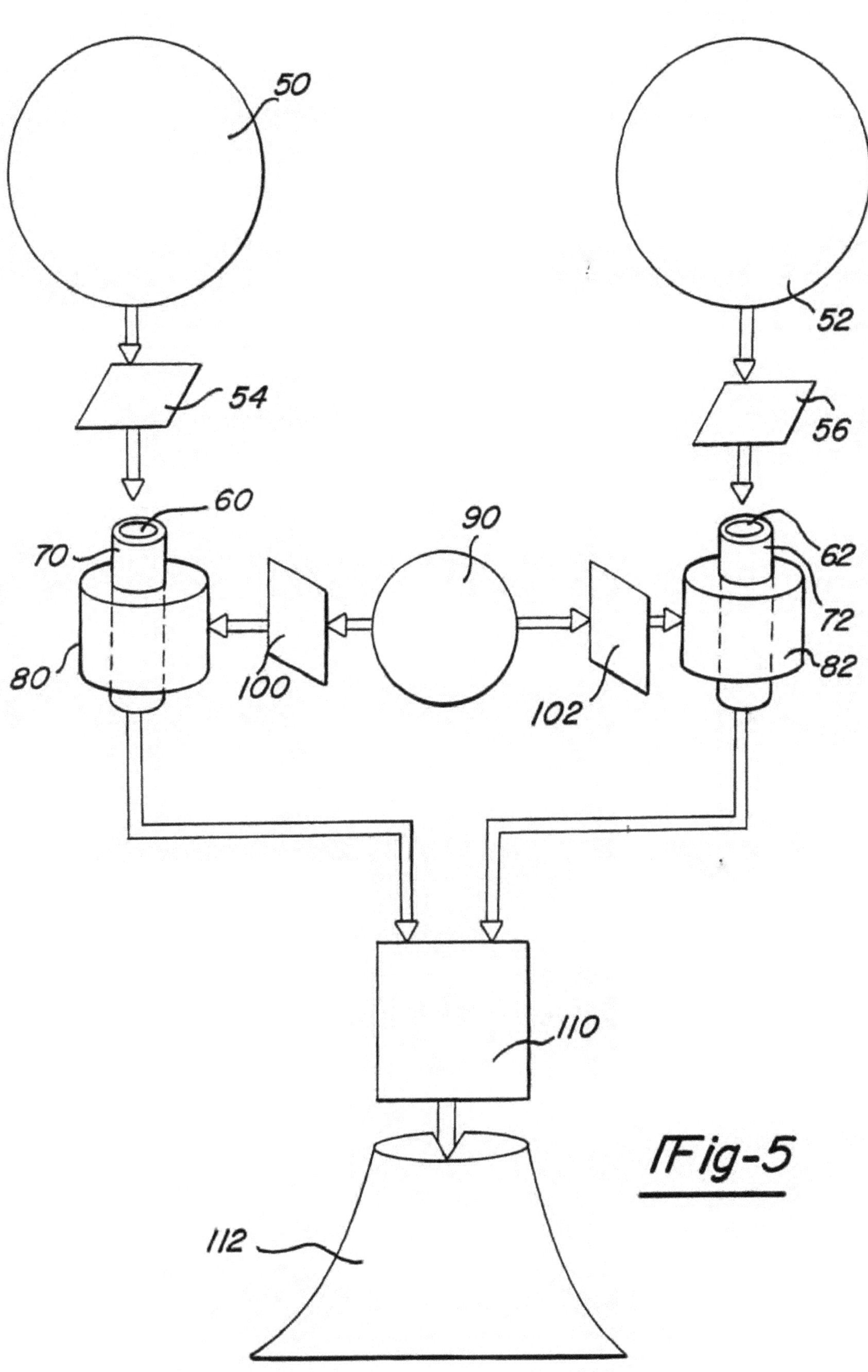

IFig-5

UK Patent Application GB 2 130 431 A

Application No **8323072**
Date of filing **26 Aug 1983**
Priority data
412548
30 Aug 1982
United States of America (US)
Application published
31 May 1984
INT CL^3
H01S 3/14
Domestic classification
H1C 397 44Y 462 48X 49X 63X 691 692 BX
Field of search
No search possible
Applicant
Meguer Vartan Kalfaian,
962 Hyperion Avenue,
Los Angeles,
California 90029,
United States of America.
Inventor
Meguer Vartan Kalfaian
Agent and/or Address for Service
Abel & Imray,
Northumberland House,
303-306 High Holborn,
London, WC1V 7LH.

Method and means for producing perpetual motion with high power

The perpetual static energies, as provided by the electron (self spin) and the permanent magnet (push and pull) are combined to form a dynamic function. Electrons emitted from a heated coil F are entrapped permanently within the central magnetic field of a cylindrical magnet M5. A second magnet M6, in opposite polarity to the poles of the electrons causes polar tilt, and precession. This precession radiates powerful electromagnetic field to a coil L between the cylindrical magnet and a vacuum chamber C - wound in a direction perpendicular to the polar axes of the electrons. Alternatively, the electromagnetic radiation is emitted as coherent light. The original source of electrons is shut off after entrappment.

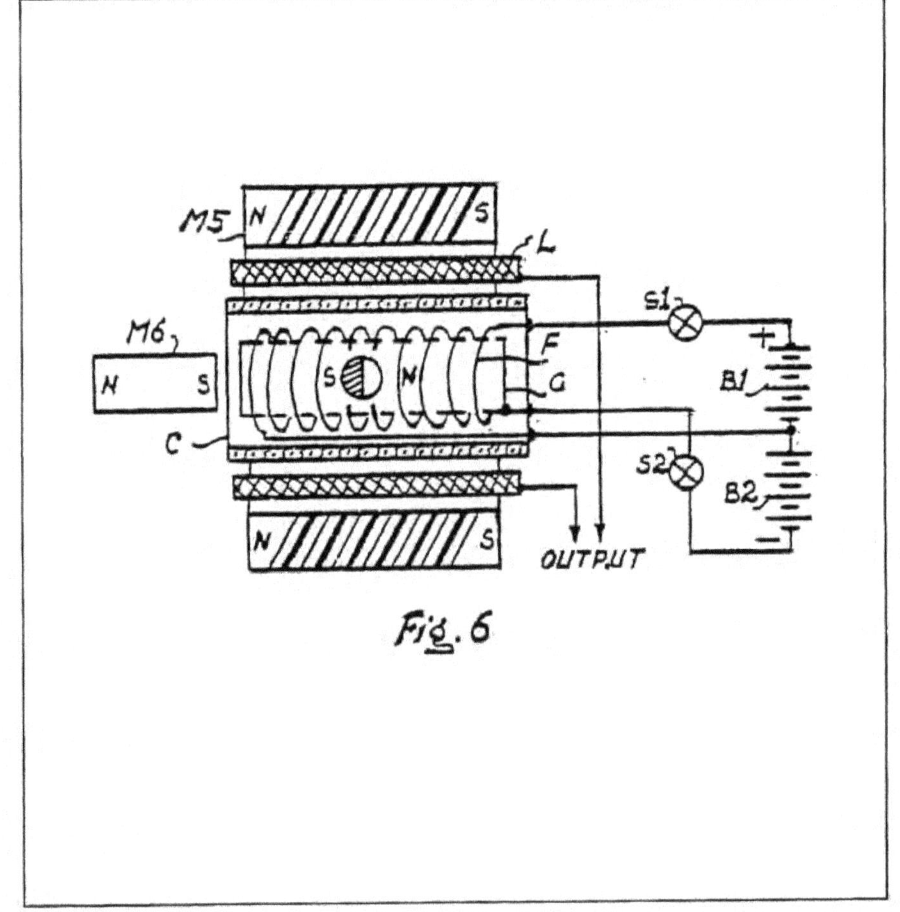

Fig. 6

GB 2 130 431 A

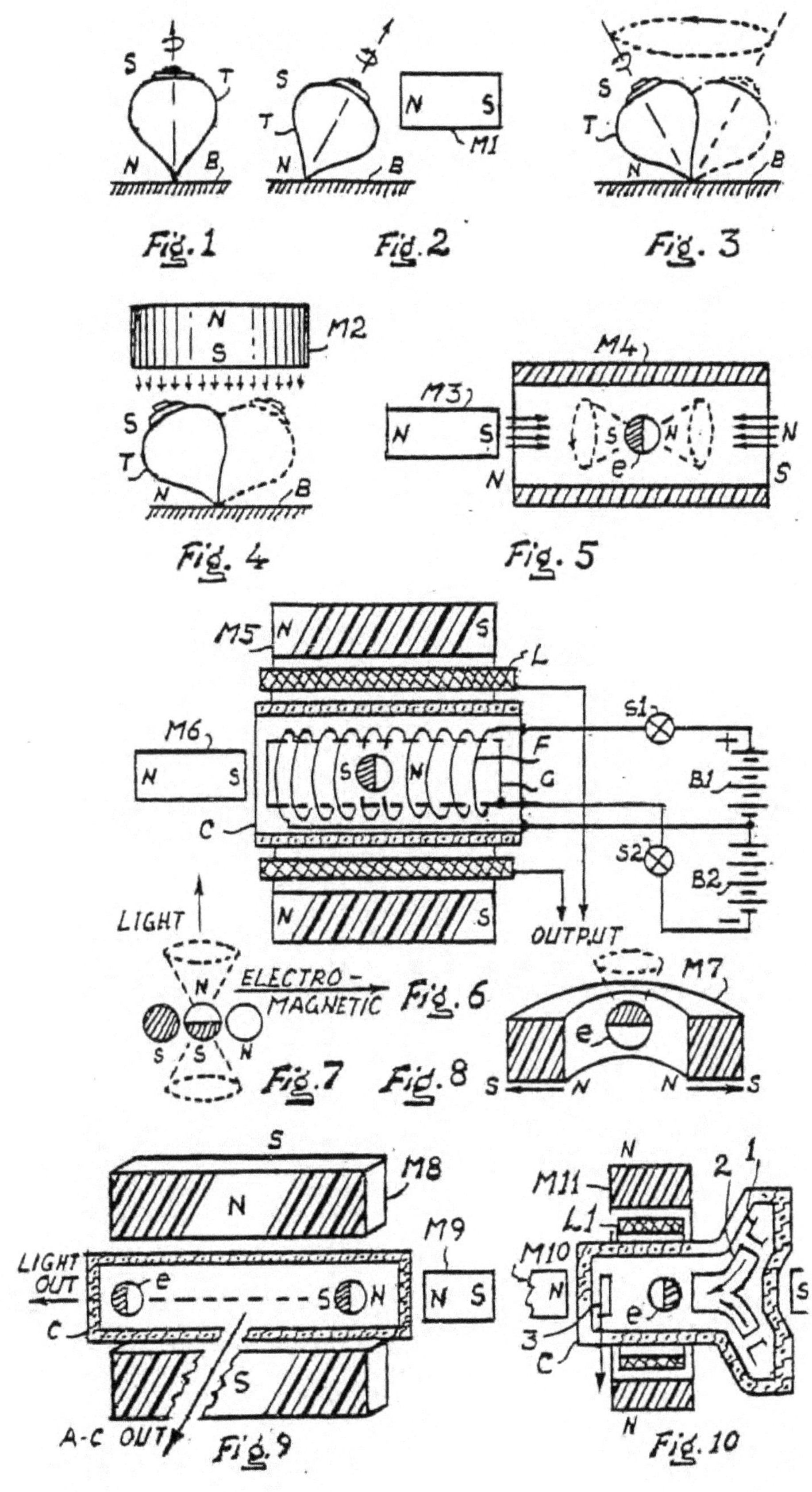
S
T
N
B
Fig. 1
S
T
N
B
N S
M1
Fig. 2
S
T
N
B
Fig. 3
N
S
M2
S
T
N
B
Fig. 4
M4
M3
N S
S N
e
N
S
N
Fig. 5
M5
N S
L
M6
N S
S N
F
G
S1
B1
C
S2
B2
N S
OUTPUT
LIGHT
N
ELECTRO-
MAGNETIC
Fig. 6
S S N
Fig. 7
Fig. 8
M7
e
S N N S
S
M8
N
M9
LIGHT
OUT
e
S N
N S
C
S
A-C OUT
Fig. 9
N
M11
L1
M10
N
e
S
2
1
3
C
N
Fig. 10

GB 2 130 431 A

SPECIFICATION

Method and means for producing perpetual motion with high power

This invention relates to methods and means for producing perpetual motion. An object of the invention is, therefore, to produce useful perpetual motion for utility purposes.

Brief embodiment of the invention

The electron has acquired self spin from the very beginning of its birth during the time of creation of matter, and represents a perpetual energy. But self spin alone, without polar motion is not functional, and therefore, useful energy cannot be derived therefrom. Similarly, the permanent magnet represents a perpetual energy, but since its poles are stationary, useful energy cannot be derived from it. However, the characteristics of these two types of static energies differ one from the other, and therefore the two types of energies can be combined in such a manner that, the combined output can be converted into perpetual polar motion.

In one exemplary mode, a cylindrical vacuum chamber having a filament and a cathode inside, is enclosed within the central magnetic field of a cylindrical permanent magnet, the magnetization of which can be in a direction either along the longitudinal axis, or from the center to the circumferential outer surface of the cylinder. When current is passed through the filament, the emitted electrons from the cathode are compressed into a beam at the center of the cylindrical chamber by the magnetic field of the cylindrical magnet. Thus, when the current through the filament is shut off, the electrons in the beam remain entrapped within said magnetic field permanently.

In such an arrangement, the poles of the electrons are aligned uniformly. When a second permanent magnet is held against the beam in repelling polarity, the poles of the electrons are pushed and tilted from their normal longitudinal polar axes. In such tilted orientations, the electrons now start wobbling (precessing) in gyroscopic motions, just like a spinning top when it is tilted to one side. The frequency of this wobbling (precessional resonance) depends upon the field strengths of the two magnets, similar to the resonance of the violin string relative to its tensional stretch. The polar movements of the electrons radiate electromagnetic field, which is receivable by an inductance for conversion into any desired type of energy. Because of the uniformly aligned electrons, the output field is coherent, and the output is high.

Observed examples upon which the invention is based

The apparatus can best be described by examples of a spinning top in wobbling motion. Thus, referring to the illustration of Figure 1, assume that the spinning top T is made of magnetic material, as indicated by the polar signs (S and N). Even though the top is magnetic, the spin motion does not radiate any type of field, for reception and conversion into a useful type of energy. This is due to the known fact that, radiation is created only when the poles of the magnet are in motion, and in this case, the poles are stationary.

When a magnet M1 is held from a direction perpendicular to the longitudinal polar axis of the top, as shown in Figure 2, the polar axis of the top will be tilted as shown, and keep on spinning in that tilted direction. When the magnet M1 is removed, however, the top will try to regain its original vertical posture, but in doing so, it will wobble in gyroscopic motion, such as shown in Figure 3. The faster the top spins, the faster the wobbling motion will be.

The reason that the top tilts angularly, but does not wobble when the magnet M1 is held from horizontal direction, is that, the one sided pull prevents the top from moving away from the magnetic field for free circular wobble. But instead of holding the magnet M1 from the side of the top, we may also hold the magnet from a direction above the top, as shown in Figure 4. In this case, however, the polar signs between the magnet and the top are oriented in like signs, so that instead of pulling action, there is pushing action between the magnet and the top - causing angular tilt of the top, such as shown in Figure 4. The pushing action of the magnetic field from above the top is now equalized within a circular area, so that the top finds freedom to wobble in gyroscopic rotation.

The important point in the above given explanation is that, the top tries to gain its original vertical position, but it is prevented to do so by the steady downward push by the static magnetic field of the magnet M2. Thus, as long as the top is spinning, it will wobble in a steady state. Since now there is polar motion in the wobbling motion of the top, this wobbling motion can easily be converted into useful energy. To make this conversion into perpetual energy, however, the top must be spinning perpetually. And nature has already provided a perpetually spinning magnetic top, which is called, the electron - guaranteed to spin forever, at a rate of 1.5 $\times 10^{23}$ (one hundred fifty thousand billion billion revolutions per second).

Brief description of the drawings

Figure 1 illustrates a magnetic spinning top, for describing the basic principles of the invention.

Figure 2 illustrates a controlled top for describing the basic principles of the invention.

Figures 3 and *4* illustrate spinning tops in wobbling states for describing the basic principles of the invention.

Figure 5 shows how an electron can be driven into wobbling state by control of permanent magnets, according to the invention.

Figure 6 is a practical arrangement for obtaining perpetual motion.

Figure 7 shows a natural atomic arrangement for obtaining precessional resonance.

Figure 8 shows a different type of electron trapping permanent magnet, as used in Figure 6.

Figure 9 is a modification of Figure 6; and

Figure 10 is a modification of the electron trapping magnets, as used in Figure 6.

Best mode of carrying out the invention

Referring to the exemplary illustration of Figure 4, the spinning top T is pivoted to the base B by gravity. In the case of the electron, however, it must be held tight between some magnetic forces. Thus, referring to the illustration of Figure 5, assume that an electron e is placed in the center of a cylindrical magnet M4. The direction of magnetization of the magnet M4, and the polar orientation of the electron e are marked in the drawing. In this case, when a permanent magnet M3 is placed at the open end of the cylindrical magnet M4, the electron e will precess, in a manner, as described by way of the spinning top. The difficulty in this arrangement is that, electrons cannot be separated in open air, and a vacuum chamber is required, as in the following:

Figure 6 shows a vacuum chamber C, which contains a cylindrically wound filament F, connected to the battery B1 by way of the switch S1. Thus, when the switch S1 is turned ON, the filament F is lighted, and it releases electrons. External of the vacuum chamber C is mounted a cylindrical permanent magnet M5, which compresses the emitted electrons into a beam at the center of the chamber. When the beam is formed, the switch is turned OFF, so that the beam of electrons is entrapped at the center of the chamber permanently.

The permanent trapping of the electrons in the hamber C represents a permanent storage of static nergy. Thus, when a permanent magnet M6 is laced to tilt the polar orientations of the uniformly oled electrons in the beam, they start processing erpetually at a resonant frequency, as determined y the field strengths of the magnets M5 and M6.

The precessing electrons in the beam will radiate uadrature phased electromagnetic field in a direcion perpendicular to the polar axes of the electrons. Thus, a coil L may be placed between the magnet M5 nd the vacuum chamber C, to receive the radiated ield from the beam. The output may then be utilized n different modes for practical purposes, for examle, rectified for d-c power use.

The electron beam-forming cylindrical magnet M5, which may also be called a focusing magnet, is shown to be bipolar along the longitudinal axis. The direction of magnetization, however, may be from the central opening to the outer periphery of the magnet, as shown by the magnet M7, in Figure 8. But the precessing magnet M6 will be needed in either case.

In the arrangement of Figure 6, I have included a current control grid G. While it is not essential for operation of the arrangement shown, it may be connected to a high negative potential B2 by the switch S2 just before switching the S1 in OFF position, so that during the cooling period of the filament, there will occur no escape of any electrons from the beam to the cathode. Also, the grid G may be switched ON during the heating period of the cathode, so that electrons are not forcibly released from the cathode during the heating period, and thereby causing no damage to the cathode, or filament.

Biological precessional resonance

Electron precessional resonance occurs in living tissue matter, as observed in laboratory tests. This is called ESR (Electron Spin Resonance) or PMR (Paramagnetic Resonance). In tissue matter, however, the precessing electron is entrapped between two electrons, as shown in Figure 7, and the polar orientations are indicated by the polar signs and shadings, for clarity of drawing.

Simulation

The arrangement of Figure 7 may be simulated artificially in a manner as shown in Figure 9, wherein, the electron trapping magnet is a pair of parallel spaced magnets M8. In actual practice, however, the structure of this pair of magnets M8 can be modified. For example, a second pair of magnets M8 may be disposed between the two pairs, so that the directions of the transverse fields between the two pairs cross mutually perpendicular at the central longitudinal axis of the vacuum chamber. The inner field radiating surfaces of these two pairs of magnets may be shaped circular, and the two pairs may be assembled, either by physical contact to each other, or separated from each other.

Modifications

Referring to the arrangements of Figures 6, 9 and 10, when the electron is in precessional gyroscopic motion, the radiated field in a direction parallel to the polar axis of the electron, is a single phased corkscrew waveform, which when precessed at light frequency, the radiation produces the effect of light. Whereas, the field in a direction perpendicular to the axis of the electron produces a quadrature phased electromagnetic radiation. Thus, instead of utilizing the output of electron precession for energy purposes, it may be utilized for field radiation of either light or electromagnetic waves, such as indicated by the arrows in Figure 9. In this case, the output will be coherent field radiation.

In reference to the arrangement of Figure 6, the electron emission is shown to occur within the central magnetic field of the focusing magnet M5. It may be practically desired, however, that these electrons are injected into the central field of the cylindrical magnet from a gun assembly, as shown in an exemplary arrangement of Figure 10. In this case, the vacuum chamber C is flanged at the right hand side, for mounting an electron emitting cathode 1 (the filament not being shown), and a curved electron-accelerating gun 2. The central part of this flange is recessed for convenience of mounting an electron-tilting magnet (as shown), as close as possible to the electron beam. In operation, when current is passed through the filament, and a positive voltage is applied (not shown) to the gun 2, the emitted electrons from the cathode are accelerated and injected into the central field of the magnet 11. Assuming that the open end of the gun 2 overlaps slightly the open end of the cylindrical central field of the magnet M1, and the positive accelerating voltage applied to the gun 2 is very low, the accelerated electrons will enter the central field of the magnet M11, and travel to the other end of the field. Due to the low speed acceleration of the electrons, however, they cannot spill out of the field, and become permanently entrapped therein.

In regard to the direction in which the coil L1 is positioned, its winding should be in a direction perpendicular to the longitudinal axis of the beam - to which the polar axes of the electrons are aligned uniformly in parallel. In one practical mode, the coil L1 may be wound in the shape of a surface winding around a tubular form fitted over the cylindrical vacuum chamber.

In regard to the operability of the apparatus as disclosed herein, the illustration in Figure 7 shows that the field output in a direction parallel to the polar axis of the electron is singular phased, and it produces the effect of light when the precessional frequency is at a light frequency. Whereas, the output in a direction perpendicular to the polar axis of the electron is quadrature phased, which is manifested in practiced electromagnetic field transmission.

In regard to experimental references, an article entitled "Magnetic Resonance at high Pressure" in the "Scientific American" by George B. Benedek, page 105 illustrates a precessing nucleus, and indicates the direction of the electromagnetic field radiation by the precessing nucleus. The same technique is also used in the medical apparatus "Nuclear magnetic resonance" now used in numerous hospitals for imaging ailing tissues (see "high Technology" Nov. Dec. 1982. Refer also to the technique of detecting Electron Spin Resonance, in which electrons (called "free radicals") are precessed by the application of external magnetic field to the tissue matter. In all of these practices, the electromagnetic field detecting coils are directed perpendicular to the polar axes of the precessing electrons or the nuclei.

In regard to the production of light by a precessing electron, in a direction parallel to the polar axis of the precessing electron, see an experimental reference entitled "Free electrons make powerful new laser" published in "high Technology" February 1983 page 69.

In regard to the aspect of producing and storing the electrons in a vacuum chamber, it is a known fact by practice that the electrons are entrapped within the central field of a cylindrical permanent magnet, and they will remain entrapped as long as the magnet remains in position.

In regard to the performance of obtaining precessional resonance of the electron, the simple example of a wobbling top is sufficient, as proof of operability.

Having described the preferred embodiments of the invention, and in view of the suggestions of numerous possibilities of modifications, adaptations, adjustments and substitutions of parts, it should be obvious to the skilled in related arts that other possibilities are within the spirit and scope of the present invention.

CLAIMS

1. The method of effecting perpetual retainment and precession of electrons, for obtaining perpetual field radiation from the polar motions of said precessing electrons, comprising the steps of:

producing electrons;

compressing said produced electrons into a perpetually retainable state; and

processing said compressed electrons for effecting perpetual field radiation by the polar motions of said precessing electrons.

2. The method of producing perpetual field radiation for conversion into perpetual energy, the method comprising the steps of:

producing electrons;

imposing a first perpetually occurring electron controlling force from a first direction upon said produced electrons into a perpetually retainable state; and

imposing a second perpetually occurring electron controlling force from a second direction upon said retained electrons, for inducing precessional motions to the electrons, and thereby obtaining said perpetual field radiation for conversion into perpetual energy.

3. The method of generating perpetual simultaneous single phased and quadrature phased coherent field radiations, comprising the steps of:

producing electrons;

imposing a first perpetually occurring electron controlling force from a first direction upon said produced electrons into a uniformly polarized perpetually retainable compressed state; and

imposing a second perpetually occurring electron controlling force from a second direction upon said compressed electrons, for effecting precessional motions of the electrons, thereby causing a quadrature phased coherent field radiation in a direction perpendicular to the uniformly polarized polar axes of said electrons, and a simultaneous single phased coherent field in a direction parallel to the polar axes of said electrons.

4. The method of producing perpetual dynamic motions for conversion into energy, comprising the steps of:

trapping and compressing a concentrated quantity of electrons within a first electron controlling field in a vacuum space, whereby forming a tightly confined permanent concentration of statistically spinning electrons, both of their polar axes and polar orientations being uniformly aligned; and

tilting the polar axes of said trapped electrons by a second permanent electron controlling field, for inducing precessional gyrations to the electrons in the form of perpetual dynamic motions, which are adaptively convertible into energy.

5. Apparatus for producing perpetual dynamic motions, which comprises:

a vacuum chamber having an electron-emitting means; an auxiliary means for causing emission of electrons from said electron-emitting means;

a first permanent magnet disposed externally of said chamber for trapping and compressing a quantity of said emitted electrons within its magnetic field, with uniform alignments of the polar axes and polar orientations of said electrons;

means for stopping said auxiliary means from

further causing emission of electrons from said electron emitting means, whereby forming a tightly confined concentration of statistically spinning electrons permanently entrapped within said first permanent magnet; and

a second permanent magnet, the field projection of which is oriented to tilt the polar a axes of said trapped electrons, for causing precessional gyrations of the electrons, as representation of said dynamic motions.

6. Apparatus comprisng:

a vacuum chamber having an electron emitting means;

an auxiliary means for causing emission of electrons from said electron emitting means;

a first permanent magnet disposed externally of said chamber for permanently trapping and compressing a quantity of said emitted electrons within its magnetic field, with uniform alignments of the polar axes and polar orientations of said electrons; and

a second permanent magnet so oriented with respect to said entrapped electrons that, the field projection from the second magnet causes precessional gyrations of the uniformly aligned entrapped electrons.

7. The apparatus as set forth in claim 6, wherein said first permanent magnet is cylindrical magnet surrounding said chamber, and the magnetization of said first magnet is in a direction along the longitudinal axis of the cylinder.

8. The apparatus as set forth in claim 6, wherein said first permanent magnet is cylindrical magnet surrounding said chamber, and the magnetization of said first magnet is in a direction from the central hollow space to the outer surface of said cylinder.

9. The apparatus as set forth in claim 6, wherein the polar sign of field projection from said second magnet to said entrapped electrons is in repelling polar sign.

10. The apparatus as set forth in claim 6, wherein is included a field responsive coil mounted between said first magnet and said vacuum chamber, for receiving the field radiation that is effected by the motions of said gyrating electrons.

11. The apparatus as set forth in claim 6, wherein is included a field responsive coil mounted between said first magnet and said vacuum chamber, the turns of winding of said coil being in a direction perpendicular to the polar axes of said compressed electrons.

12. Apparatus for producing perpetual motion, the apparatus being substantially as hereinbefore described with reference to, and as illustrated by, the accompanying drawings.

Printed for Her Majesty's Stationery Office, by Croydon Printing Company Limited, Croydon, Surrey, 1984.
Published by The Patent Office, 25 Southampton Buildings, London, WC2A 1AY, from which copies may be obtained.

QUESTIONS & ANSWERS WITH THE AUTHOR

Q: DO I NEED TO BUY ALL SIX BOOKS TO BE ABLE TO BUILD A UFO CRAFT?

A: NO. READING THE FULL SERIES WILL INCREASE YOUR OVERALL KNOWLEDGE, BUT THE KNOWLEDGE TO BUILD INDIVIDUAL AND UNIQUE UFO CRAFT IS CONTAINED WITHIN EACH INDIVIDUAL VOLUME.

Q: HOW BIG ARE THEY? (ONE, TWO, FOUR PASSENGER?)

A: SIZES ARE LIMITED ONLY BY THE DESIGNS OF THE BUILDER.

Q: WHAT WOULD BE THE COST TO BUILD MY OWN?

A: COSTS WILL VARY ACCORDING TO TYPE.

FIRST MUST BE DETERMINED THE PRIMARY PROPULSION SYSTEM, FOLLOWED BY BACK-UP SYSTEMS.

1) ELECTROGRAVITIC
2) AIR FOIL
3) PLASMA THRUST

4) MAGNETOHYDRODYNAMIC
5) INERTIAL PROPULSION

COSTS WILL VARY DEPENDING ON DESIGN FACTORS, AVAILABILITY OF MATERIALS IN YOUR AREA, WHAT FABRICATION REQUIREMENTS WILL BE OUTSOURCED...

Q: WOULD THEY BE DIFFICULT TO BUILD AT HOME?

A: THAT ANSWER DEPENDS ENTIRELY ON THE SKILLS AND MACHINING FACILITIES AVAILABLE TO THE INDIVIDUAL BUILDER. THE REQUISITE SKILLS WOULD INCLUDE MACHINING, WORKING WITH CERAMICS, ELECTRICAL WIRING SKILLS, AND GENERAL ENGINEERING. ANY PLANT THAT MANUFACTURES AIRPLANES, BOATS, OR AUTOMOBILES COULD BE CONVERTED TO BUILD UFO CRAFT. ANYONE THAT KNOWS HOW TO BUILD A TESLA COIL HAS A GOOD START ON HOW TO BUILD UFO PROPULSION SYSTEM APPARATUS.

Q: DO YOU PROVIDE CONSULTATION SERVICES TO BUILD?

A: NO. AT THIS POINT IN TIME, I AM NOT AN ENGINEER, MACHINIST, NOR METAL-WORKER.

Q. DO YOU OFFER ANY CONSTRUCTION TIPS?

A: NO. THE TECHNOLOGY TO BUILD UFO CRAFT INVOLVES HIGH VOLTAGES, POWERFUL ELECTROMAGNETIC FIELDS, HIGH TEMPERATURES, POTENTIAL RADIOACTIVITY[1], AND/OR HIGH-SPEED ROTATIONS THAT ARE POTENTIALLY DANGEROUS TO LIFE AND PROPERTY. IF YOU ARE NOT FAMILIAR WITH THE SAFE PRACTICES INVOLVED IN WORKING WITH SUCH METHODS, MATERIALS, AND CIRCUMSTANCES, THEN YOU SHOULD NOT ATTEMPT SUCH CONSTRUCTION WITHOUT ASSEMBLING A QUALIFIED, LICENSED, INSURED, AND BONDED TEAM OF PROFESSIONALS WITH SUCH EXPERIENCE. SAFETY FIRST!

[1] METALS EXPOSED TO EXTREME VOLTAGES AND POWERFUL EMF CAN BECOME RADIOACTIVE, AS ANYONE WHO HAS BUILT ONE OF PHILO FARNSWORTH'S FUSORS CAN TESTIFY TO. FOR MORE EVIDENCE SEE GUSTAVE LE BON'S "THE EVOLUTION OF MATTER" OR JUST GOOGLE "INDUCED RADIOACTIVITY".

Q: WOULD THEY BE LEGAL TO FLY?

A: DEPENDING ON WHAT COUNTRY YOU ARE IN, THERE MORE THAN LIKELY EXISTS A GOVERNMENTAL DEPARTMENT THAT CAN PROVIDE YOU THE NECESSARY DOCUMENTATION TO INSURE THE LEGALITY OF YOUR CRAFT AND FLIGHT PLANS.

Q: IS INTERPLANETARY TRAVEL A POSSIBILITY WITH THESE TECHNOLOGIES?

A: ABSOLUTELY. THE OPENING PATENT IN VOLUME III, "PLASMA PROPULSION", A 1950'S PATENT HAS A THRUST 60 TIMES MORE POWERFUL THAN CONVENTIONAL ROCKETS. ONE OF THE PATENTS IN THE SECOND VOLUME, "ELECTROGRAVITICS", HAS INITIAL VELOCITY CALCULATED BY ONE RESEARCHER TO BE IN EXCESS OF 25,000 MPH.

T. TOWNSEND BROWN, IN MANY WAYS THE FATHER OF ELECTROGRAVITICS, BELIEVED THAT ELECTROGRAVITIC PROPULSION IN A VACUUM COULD EXCEED LIGHT SPEED.

Q: HOW DO YOU PROTECT HUMANS FROM THE HARSHNESS OF SPACE, LIKE COSMIC RADIATION, MAGNETIC BELTS OF PLANETS, ETC.?

A: A MAGNETOHYDRODYNAMIC PLASMA SHELL GENERATED BY THE CRAFT WOULD, IN THEORY, INSULATE AGAINST HARMFUL COSMIC RADIATION. SUCH A PLASMA SHELL IS GENERATED BY THE CRAFT AT THE BEGINNING OF THIS BOOK AND IN VOLUME IV "MAGNETOHYDRODYNAMICS".

Q: CAN WE SEE YOUR PERSONAL PROGRESS WITH BUILDING A UFO?

A: THAT INFORMATION IS ACCESS RESTRICTED TO SIGMA SQUAD MEMBERS.

Q: HOW DOES SOMEONE BECOME A SIGMA SQUAD MEMBER?

A: YOU CAN JOIN THE SIGMA SQUAD BY SIGNING UP FOR THE SIGMA SQUAD NEWS BLOG AT WWW.UFOHOWTO.COM/SIGMA - SQUAD.HTM.

Q: WHAT DO YOU THINK OF THE GROWING BODY OF ANCIENT UFO EVIDENCE THAT SHOWS UP IN ANCIENT RELIGIOUS TEXTS AND ARTWORK?

A: THE WHOLE ANCIENT ALIEN-DIVINITY THING IS A LITTLE BEYOND THE SCOPE OF MY BOOKS. OTHER AUTHORS HAVE WRITTEN ABOUT THAT, AND I KNOW THAT THE VIMANA CRAFT "BLUEPRINTS" ARE AVAILABLE ONLINE.
I WOULD LOVE TO SEE THE HUMAN RACE MOVE INTO A FUTURE THAT INVOLVED A GALACTIC OR COSMIC SOCIETY.
I WANT TO MAKE SUCH KNOWLEDGE AVAILABLE, THAT THE ENTERPRISING KNOW WHERE TO LOOK, TO BUILD THEIR OWN CRAFTS, LIVES, AND FUTURES.

Q: ISN'T THERE A GOVERNMENT CONSPIRACY TO CONCEAL THIS KNOWLEDGE FROM THE PUBLIC?

A: ALL THE INFORMATION CONTAINED IN THE UFO HOW-TO SERIES COMES FROM A GOVERNMENT DEPARTMENT: THE PATENT OFFICE.

Q: BUT WHAT ABOUT AREA 51, AND MILITARY BASES THAT THREATEN PEOPLE THAT SEE WHAT THEY'VE BEEN WORKING ON?

A: THE MILITARY WILL PROTECT ITS RESEARCH IN THE INTEREST OF NATIONAL SECURITY, IN THE SAME MANNER THAT INTELLIGENCE AGENCIES WILL DO SO. BUT JUST BECAUSE RESEARCH IS BEING DONE IN A FIELD DOES NOT MEAN THAT FIELD IS OFF LIMITS TO EVERYONE ELSE. THE MILITARY DOES RESEARCH WITH BALLISTICS, GAS POWERED ENGINES, FUEL CELLS... BUT THAT TECHNOLOGY IS NOT FORBIDDEN FROM THE PUBLIC. THE RESULTS OF THE PARTICULAR RESEARCH MAY BE CLASSIFIED, BUT THE FIELDS ARE OPEN. SAME THING WITH UFO TECHNOLOGY. THE PATENTS HAVE EXISTED IN THE PUBLIC DOMAIN FOR 100 YEARS.

Q: SO WHY DON'T MORE PEOPLE KNOW ABOUT THIS STUFF?

A: THERE ARE A NUMBER OF DIFFERENT THEORIES THAT PEOPLE HAVE AS TO WHY THE KNOWLEDGE OF THESE INVENTIONS HASN'T BEEN MADE COMMONPLACE. I

DEFER TO ANCIENT ROMAN WISDOM. "QUI BONO?" WOULD BE ASKED, TO DETERMINE WHO WAS MOST LIKELY RESPONSIBLE FOR A WRONG-DOING. "WHO BENEFITS?" IN THIS CASE, WHO BENEFITS FROM THE SUPPRESSION OF TECHNOLOGY THAT COULD MAKE AIRLINES, TRANS-OCEANIC SHIPPING, AND GASOLINE POWERED VEHICLES OBSOLETE?

Q: WHAT IMPACT WILL “THE UFO HOW-TO” SERIES HAVE ON SPACE TOURISM?

A: SPACE TOURISM WILL NO LONGER BE A FIELD DOMINATED BY WEALTHY PERSONS AND POWERFUL CORPORATIONS. WITH THE TECHNOLOGY FINALLY IN REACH OF THE AVERAGE PERSON, THE FINAL FRONTIER WILL BE THE NEXT GOLD RUSH; SPACE TOURISM WILL BE THE NEXT GREAT WESTWARD EXPANSION.

I SEE SPACE TOURISM AS NOT CONCEIVED BEFORE, HONEYMOON TOURS TO THE MOON, MONTH-LONG CRUISES AROUND THE OUTER GIANTS, SHOOTING GALLERIES IN THE ASTEROID BELT, AND MANKIND EXPANDING THROUGH THE SOLAR SYSTEM.

Q: WHAT IS THE BEST WAY FOR PEOPLE TO CONTACT YOU?

A: E-MAIL! PLEASE SEND ALL QUESTIONS AND COMMENTS TO:

AUTHOR@UFOHOWTO.COM

LET'S ADVANCE INTO THE FUTURE TOGETHER!

The Titles of the UFO How-To Aerospace Technical Manuals available at Lulu.com:

100 Years of UFO Patents
http://www.lulu.com/shop/luke-fortune/ufo-how-to-aerospace-technical-manual-volume-i-100-years-of-ufo-patents/paperback/product-21108273.html

Electrogravitics
http://www.lulu.com/shop/luke-fortune/ufo-how-to-aerospace-technical-manual-volume-ii-electrogravitics/paperback/product-21108408.html

Plasma Propulsion Systems
http://www.lulu.com/shop/luke-fortune/ufo-how-to-aerospace-technical-manual-volume-iii-plasma-propulsion-systems/paperback/product-21118627.html

Magnetohydrodynamics
http://www.lulu.com/shop/luke-fortune/ufo-how-to-aerospace-technical-manual-volume-iv-magnetohydrodynamics/paperback/product-21119923.html

Fusion and Antimatter Systems
http://www.lulu.com/shop/luke-fortune/ufo-how-to-aerospace-technical-manual-volume-v-fusion-and-antimatter-systems/paperback/product-21121259.html

Inertial Propulsion Systems
http://www.lulu.com/shop/luke-fortune/ufo-how-to-aerospace-technical-manual-volume-vi-inertial-propulsion-systems/paperback/product-21122325.html

Esoteric Power Systems
http://www.lulu.com/shop/luke-fortune/ufo-how-to-aerospace-technical-manual-volume-vii-esoteric-power-systems/paperback/product-21122633.html

Permanent Magnet Power Systems
http://www.lulu.com/shop/luke-fortune/ufo-how-to-aerospace-technical-manual-volume-viii-permanent-magnet-power-systems/paperback/product-21122746.html

Hydrogen Power Systems
http://www.lulu.com/shop/luke-fortune/ufo-how-to-aerospace-technical-manual-volume-ix-hydrogen-power-systems/paperback/product-21123451.html

EM UFO Systems
http://www.lulu.com/shop/luke-fortune/ufo-how-to-aerospace-technical-manual-volume-x-em-ufo-systems/paperback/product-21123679.html

NAV/COM Systems
http://www.lulu.com/shop/luke-fortune/ufo-how-to-aerospace-technical-manual-volume-xi-navcom-systems/paperback/product-21125215.html

On The Threshold Of Tomorrow
http://www.lulu.com/shop/luke-fortune/ufo-how-to-aerospace-technical-manual-volume-xii-on-the-threshold-of-tomorrow/paperback/product-21125445.html

www.ingramcontent.com/pod-product-compliance
Ingram Content Group UK Ltd.
Pitfield, Milton Keynes, MK11 3LW, UK
UKHW061830190726
13855UKWH00005B/1740

9 781312 219014